THE OMNI BOOK
OF COMPUTERS & ROBOTS

Prepare yourself for *Omni*'s vision of life to come.
But be warned: The bad is here as well as the good,
computerized war and electronic crime as well as
robot servants and silicon art. Here you'll find a
future that is fast becoming a reality—a reality of
wonders . . . .

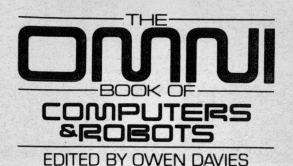

THE OMNI
BOOK OF
COMPUTERS & ROBOTS

EDITED BY OWEN DAVIES

ZEBRA BOOKS
KENSINGTON PUBLISHING CORP.

ZEBRA BOOKS

are published by

KENSINGTON PUBLISHING CORP.
475 Park Avenue South
New York, N.Y. 10016

Omni is a registered trademark of Omni Publications International, Ltd.

CONTENTS

5

PART FOUR: LIVEWARE

PART FIVE: COMPUTERS THAT WALK

PART SIX: DIGITS AND DOLLARS

PART SEVEN: CYBERCONFLICTS

PART EIGHT: AUTOMATED ARTS

PART NINE: A SKEPTICAL EYE

FOREWORD

There have been four great moments in the computer revolution. Only one has meant anything to the average man:

In 1821, the great English mathematician Charles Babbage announced that he would build a "Different Engine," a complex mass of gears and cogs that would calculate the answers to algebraic equations. His scheme included all the essential parts of a modern computer. It failed because craftsmen then could not machine parts accurately enough to build a working model. In the entire world, only a few hundred people ever understood what Babbage had attempted.

In 1947, scientists at Maryland's Moore School of Engineering switched on their Electronic Numerical Integrator and Computer—ENIAC, the first general-purpose electronic computer ever built. At the time, one specialist estimated that it would take only four such marvels to handle all the computation the world required.

In 1971, a then-small company called Intel announced an electronic breakthrough—the first microprocessors, circuits that condensed an entire computer onto a single chip of silicon. They were a marketing flop until one of them found its way into the first personal computer. Soon, there were microprocessors by the millions. Yet computers remained a mystery to all but a tiny minority.

Then, in October 1978, *Omni* arrived to explain what it all meant. Not a moment too soon, for the revolution was already in full swing. Computerized bank tellers were gobbling our paychecks, usually even crediting us for them on the next statement. Home-computer stores were launching a billion-dollar business. Robots were displacing human workers from auto assembly lines. And only a few specialists could foresee where the process would lead.

It was clear from the first that in *Omni* the computer age had finally found its voice. For *Omni* was not merely a beautiful magazine; it knew the meaning of science and technology. *Omni*'s writers understood that the computer would change our world, and they had the skill to explain electronicized life to others—without putting people to sleep or burying them under strange jargon.

Omni staked out its field quickly. In its second issue, Ted Nelson offered "Computer Lib," a breezy demystification of the new technology. Soon after, Johnathan V. Post looked at "Cybernetic War"; battles of the future, he found, will be fought not so much by infantrymen with rifles as by planners using computers to analyze intelligence from digital spy satellites and operating robot weapons. And in *Omni*'s first anniversary issue, Thomas Hoover's "Intelligent Machines" explored the difficulty of making computers that think as human beings do.

In all these ground-breaking articles, and the many that have followed them, *Omni*'s goal has been clear: to explore the future in which we all must live. There were no cryptic circuit diagrams, no listings of arcane code for fevered "hackers" to punch into pizza-smeared terminals during midnight programming binges. Instead there was good writing, and behind it a sure knowledge that the future will be an exciting place to live.

Some of the best of that writing is collected in this anthology. Here you will find *Omni*'s vision of life to come. The bad is here as well as the good, computerized war and electronic crime as well as robot servants and silicon art. The twenty-eight articles here are a look at your own future, a future that is fast becoming reality. Where fair warning is needed, they give it. More often they grant an exciting foretaste of wonders that will soon enrich us all. In either case, they are never dull. They always entertain.

—Owen Davies

PART ONE:
VIEWPOINTS

COMPUTER LIB

By Ted Nelson

Suppose you met a man who didn't know what an automobile was. You'd think he was a cluck, a simpleton, someone unable to hold a sensible opinion on anything in the modern world. And it's so simple—*an automobile is a box you go places in*.

The majority of well-educated Americans, however, are clucks and simpletons when it comes to something just as basic as the automobile: the computer. Just as the automobile is a box you go places in, *the computer is a box that follows a plan*. The plan can be anything. It can tell the box to turn things on and off, to file and bring back information automatically, to blink lights, to mix drinks, to play music—anything at all.

So I've given you the word. You now know what a computer is.

* * *

They should never have been called computers.

The first machines were labeled computers because their developers felt that numerical computations would be their main function. It reminds one of the blind men in the fable, who when confronted with an elephant thought it was a wall, a tree, or a snake, depending on what parts of the animal their hands touched.

The pioneers were not wrong, just preoccupied. But the result has been that the name *computer* has frightened people ever since. It might just as well have been called the Oogabooga Box. That way at least we could get the fear out in the open and laugh at it.

In France they call all aspects of computer use *l'informatique,* the automatic handling of information. From time to time computer fans have advocated translating the French term and calling the use of computers *informatics,* an extremely appropriate term: Unfortunately, "informatics" happens to be the trademark of Informatics, Inc. here in the United States, so that pretty much squashes that.

In Sweden they call computers "dators." A computer—or a terminal between you and the computer—is a dator, meaning something that handles data. Straightforward, huh?

The celebrated scientist John Von Neumann got it right at the very beginning, but nobody listened. He called it "the all-purpose machine."

But computer is what we call it, even when it is playing music or making pictures on a screen.

Keeps people scared. Oogabooga!

* * *

Any nitwit can understand computers.
Many do.

Many people, particularly those who call themselves "humanists," often claim that computers
are oppressive, cold, impersonal, rigid, dictatorial,
militaristic—the list is endless. "They're controlling our lives!" I hear. "They're taking over the
world!" This is a widespread view, held fanatically
by a large number of people, and it is not altogether wrong.

Computers have been used in many systems that
push people around. They are frequently instruments of oppression. The Nazis, however, used
railroad cars and ovens in oppressive ways, but this
does not mean that railroad cars and ovens are in
themselves oppressive. Computers, in fact, have
provided a handy excuse for oppression: "Sorry
about this, but the computer made me do it."

Consider this: Suppose someone creates a system
for handling warranty repairs. You go into the
store where you bought your nice new radio that
doesn't work. You have the sales slip and your
registration card stub.

"Fill this out," the clerk says. He hands you a
form that asks for your name, age, sex, address,
occupation, height, weight, identifying marks,
where you purchased the appliance, where you first
heard of the appliance, a description of your
problem, and your signature.

17

You demand to know why you have to answer so many questions.

"Fill it out, or you don't get your radio fixed," responds the clerk. "Gotta have all that for the computer."

You fill out the form. It takes 20 minutes. The ballpoint pen smears ink all over your hands.

"This is no good," says the clerk. "Machine can't read the carbon copy."

Quivering with rage, you fill out another form, which the clerk accepts. You get a portion of it back that bears the first three letters of your name, several digits, and some hyphens. You hand over your radio.

"When will it be ready?" you ask.

"Six weeks, maybe seven," he answers. "Computer has to match it up with your warranty card in California."

"Never mind, just give it back," you say.

The clerk hands you your radio and confiscates the stub.

"Damn these computers," you shout and stomp out the door. For days you complain to everyone about "those damned computers never work."

But you are wrong. The system worked perfectly. They did not have to repair your radio.

And you blamed the computer.

Suppose you were taken on a tour of a library, but you did not know how to read. You would see people moving books around, turning pages. You would think they were all doing the same thing.

18

You would not see the romance and adventure, the science and history, moving across their minds.

Similarly, people working with computers all seem to be doing the same things—typing odd messages into the machine and scratching their heads at the results, staring at walls, scribbling strange symbols on blackboards, sometimes muttering out loud and walking into things.

But each one is doing something different. One may be studying the life and death of stars and galaxies, another arranging a dating service. One may be robbing a bank.

To understand the language of automobiles is basic. "A guy in a pickup tailgated me off the interstate." Everybody understands that.

To understand the language of books is basic. "The librarian told me the book was out for rebinding." Everybody understands that.

But how about this one: "The program got hung in a loop and I tried to do a restart, but the system bombed."

A few years from now, that language and the world it describes will be as familiar to us as the librarian and the tailgater are. Today, however, anyone who understands that mysterious sentence is called a "computer person."

"You'll have to talk to my nephew; he uses computers." I gnash my teeth at that statement. Just because I use wheels, must I talk to the niece who rollerskates or to the uncle who teaches driving? Luckily, in a few years there will not be any computer people. That is, people won't be set apart as computer people, any more than skateboarders and truckers and commuters are set apart as "wheel people."

This is the magic time. It is like the Klondike, like the old Hollywood, like the birth of air travel and radio and television. It is the new computer world. It is going to explode.

Today real computers can be purchased for as little as $600 (the TRS-80 from Radio Shack, for example), and some of these small machines are more powerful than the IBM 1401, the computer that ran American business only 15 years ago (and cost upwards of $50,000). Though only half a dozen brands are available at this writing, new electronics companies are tumbling into the personal computer field willy-nilly.

In my recent book, *The Home Computer Revolution* (available for $2.00 plus postage from: The Distributors, 702 So. Michigan, South Bend, IN 46618; softcover, 224 pages), I predicted that ten million small computers would be sold by the end of 1979. That prediction will probably not come true for there is not now the production capacity. But I think it will be closer to ten million than to one million.

The history of electronics since World War II can be described in two words—smaller and cheaper. The vacuum tubes that made the first hi-fi sets hum in the early 1950s were the size of a pinecone. The transistors that replaced them were first the size of a thimble, then as tiny as a BB.

Not only were transistors small—they could be grown, like little mushrooms, and sculpted and arranged in patterns while they were growing. This led to the "integrated circuit," a little cluster of transistors and other things electronic all grown together in their intended combination. The size of a postage stamp, the integrated circuit could be designed for any electronic purpose—as an amplifier for a hi-fi, a circuit for controlling radar sweeps, or the building block of a computer.

By 1965 it became clear that the workings of an entire computer could be put on a single integrated circuit. Did people plan ahead? Did the computer industry prepare for drastic change as the price of computers fell from thousands to hundreds of dollars? Did any big companies get ready for this change?

You bet they didn't.

The breakthrough came in 1971, when an integrated circuit company, Intel Corporation, brought out the first "computer on a chip," the model 4004. This was not a full computer, with blinking lights and memory, but it was the vital part—the circuitry capable of following a program stored in whatever memory that was attached to it.

Intel also offered, to go with the 4004, components that would provide memory, outside hookup, and so on. All you had to do was buy the various parts and know exactly what you were doing.

Then came the Altair. Out of the blue, in December of 1974, there came an electrifying announcement from Albuquerque, New Mexico. A tiny company called MITS was offering a computer kit—for $420.

Two hundred orders for the Altair would allow MITS, faltering, nearly broke, to break even. Those 200 orders came the day the kit came out. Quickly there were thousands of orders, then pandemonium. This tiny firm had discovered by dumb luck what a few prophets had claimed but what nobody believed: *people want computers.*

By June of 1974 there were several companies making Altair accessories. There was also a slick magazine for computer hobbyists called *Byte* and a store in Los Angeles where you could walk in and buy a computer. By December there were more than a dozen companies making Altair add-on products. By the following summer the Los Angeles computer hobby club had 3000 members.

Today there are perhaps 50 brands of personal computers on the market, most of them kits but some, like the Apple II and Radio Shack's TRS-80, fully constructed and ready to run. The revolution has begun. Computer Lib has become a fact.

Virtually every city in the country now has its own band of computer hobbyists. The weekly meeting of personal computer enthusiasts in such cities as Boston, Los Angeles, and San Francisco may draw more than 1000 members. Curiously Women's Lib has not joined up with Computer

Lib. Membership in personal computer organizations is almost exclusively male. Women are not only welcome but encouraged to join, for most male computer hobbyists are dying to meet a woman who can talk about computers. Women should beware, however, for many computer hobbyists cannot talk about anything else. "What else is there?" is the common response.

Anyone can learn to program a computer. It is simply a matter of getting access to a computer and learning one of the languages that will direct it. You don't have to know mathematics or electronics any more than you have to know the foxtrot.

The magic age, however, seems to be 14. (The average age of people in noncredit computer courses, for example, is 14.) Today, of course, there is a simple way for a youngster to get involved with computers, probably the best way of all. *His father can buy him one.*

A curious strategy pervades the computer world from top to bottom. It is employed by the grandest bureaucrats and the most modest individuals. It is known as *lock-in*.

Lock-in simply means keeping someone a prisoner of your products and services. It has been around for years—ever since Samuel Colt invented the gun you had to buy refills for. Anything that requires refills made by the manufacturer locks you in—razor, camera.

Computers and their hardware are different from manufacturer to manufacturer. This assures that after programs are written and corrected until they work perfectly, you can't change computers. Although IBM is the most notorious user of this strategy, all computer manufacturers practice lock-in whenever they can.

But lock-in is not limited to the manufacturer of computers. It also is practiced by the so-called "computer center," the department within a company, university, or government where the big computers live. Initially designed to provide centralized, efficient computer service, the computer center has evolved into an internal tyranny, set up to operate at its own convenience and dedicated (like any organizational entity) to its own self-preservation. The computer center, of course, always wants a better (read bigger) computer over which it has exclusive control.

Today, thanks to small computers that are cheap enough to bypass the "computer selection committee," the internal monopoly of the computer center is coming apart at the seams. According to Portia Isaacson, part owner of Houston's Micro Store, little computers have already penetrated large companies without the knowledge of the computer centers. The Trojan horse is already inside the gates. "I'll sell you a computer under any name you like," says Isaacson. "You'd be surprised at all the different things we call them on the sales slips."

Lock-in also is a strategy employed by individuals—programmers and technicians, primarily. If you are the only one who understands the computer system you've created, you can't be fired. All you have to do is look more and more harried each day, keep longer and longer hours, and demand raise after raise. Said one programmer: "I always tell them that if it can't be done in COBOL (the standard business-programming language) it can't be done by computer. It saves me an awful lot of trouble." The case was well made by Robert Townsend in his best-selling book, *Up the Organization*. "Most of the computer technicians you're likely to meet or hire are complicators, not simplifiers," he said. "They're trying to make it look tough, not easy."

This is a critical time for the home or personal computer. In the very near future virtually every device or appliance costing over $50 will contain computer *chips,* but it is unlikely that we will be able to treat them as computers. Each device will be programmed to behave in a specific way when you touch its buttons, but there will be no tie-in. (Sometimes this situation is called "distributed intelligence.") For example, the automatic "trip computer," already available in Cadillac, has its own fixed repertory of behaviors as does this portable telephone-memorizer (now available for about $70) and the box you can preload with your appointments (at several hundred dollars). None of them can be together. Each one locks you in.

This is not Computer Liberation. It is just

crowding us with more gadgets.

Only if the personal computer can perform a unifying function, only if it can both keep records and orchestrate our accessories, can we derive full benefit. If all we get is a lot of separate gizmos the computer revolution will fail.

Unification is not easy. It requires deep and thoughtful design. Unfortunately, so long as people buy cameras they can't understand and hi-fi sets with rows of identical knobs and switches that cannot be distinguished by touch, there will be little improvement. The computer manufacturers will build what people will buy. It is up to you.

ELECTRONIC TUTORS

By Arthur C. Clarke

We are now witnessing one of the swiftest and most momentous revolutions in the entire history of technology. For more than a century the slide rule was the essential tool of engineers, scientists, and anyone else whose work involved extensive calculations. Then, just a decade ago, the invention of the pocket calculator made the slide rule obsolete almost overnight, and with it whole libraries of logarithmic and trigonometric tables.

There has never been so stupendous an advance in so short a time. Simply no comparison can be made between the two devices. The pocket calculator is millions of times more accurate and scores of times swifter than the slide rule, and it now actually costs less. It's as if we'd jumped overnight from bullock carts to the Concorde—and Concorde were cheaper! No wonder the slide rule manufacturers have gone out of business. If you have a good slide rule, leave it in your will. It will someday be a valuable antique.

Pocket calculators are already having a profound effect on the teaching of mathematics, even at the level of elementary arithmetic. But they are about to be succeeded by devices of much greater power and sophistication—machines that may change the very nature of the educational system.

The great development in our near future is the portable electronic library—a library not only of books, but of films and music. It will be about the size of an average book and will probably open in the same way. One half will be the screen, with high-definition, full-color display. The other will be a keyboard, much like one of today's computer consoles, with the full alphabet, digits, basic mathematical functions, and a large number of special keys—perhaps 100 keys in all. It won't be as small as some of today's midget calculators, which have to be operated with toothpicks.

In theory, such a device could have enough memory to hold all the books in the world. But we may settle for something more modest, like the *Encyclopedia Britannica,* the *Oxford English Dictionary,* and *Roget's Thesaurus.* (Incidentally, Peter Mark Roget was the inventor of the log-log slide rule.) Whole additional libraries, stored in small, plug-in memory modules, could be inserted into the portable library when necessary. All this technology has already been developed, though for other uses. Oddly enough, the most skilled practitioners of this new art are the designers of video games.

Reading material may be displayed as a fixed page or else "scrolled" so that it rolls upward at a comfortable reading rate. Pictures could appear as in an ordinary book, but they may eventually be displayed as three-dimensional holographic images. Text and imagery, of course, will be accompanied by sound. Today's tape recorders can reproduce an entire symphony on a cassette. The electronic library may be able to play back the complete works of Beethoven while displaying the scores on its screen.

And how nice to be able to summon up Lord Clark's *Civilisation* or Jacob Bronowski's *Ascent of Man* whenever or wherever you felt like it! (Yes, I know that these tapes currently cost about $2,000 apiece, unless you are lucky or wicked enough to have a pirated copy. But one day the BBC will get its money back, and thereafter the price will be peanuts.)

I still haven't touched on the real potential of this technology, the opportunity to cure one of the great failings in conventional education, especially in large classes. Genuine education requires feedback—interaction between pupil and teacher. At the very least, this allows the student to clear up points he does not understand. Ideally, it provides inspiration as well. Yet I recently met a Turkish engineer who said that all he had ever seen of his professor was the tiny figure up on the platform, above a sea of heads. It is a predicament shared by all too many students.

The electronic tutor will go a long way toward solving this problem. Some computer programs already allow the student to carry on a dialogue with the computer, asking it questions and answering the questions it asks. "Computer-aided instruction"—CAI, not to be confused with CIA—can be extremely effective. At best, the pupil may refuse to believe that he is dealing with a computer program and not with another human being.

Technology's influence on education is nothing new. There's an old saying that the best educational setup consists of a log with teacher at one end and pupil at the other. Our modern world is not only woefully short of teachers, it's running out of logs. But there has always been a shortage of teachers, and technology has always been used to ameliorate this—a fact that many people tend to forget.

The first great technological aid to education was the book. You don't have to clone teachers to multiply them. The printing press did just that, and the mightiest of all educational machines is the library. Yet this potent resource is now about to be surpassed by an even more remarkable one, a depository of knowledge as astonishing to most of us today as books were to our remote ancestors.

I can still recall my own amazement when, at a NASA conference less than ten years ago, I saw my first "electronic slide rule." It was a prototype of the HP 35, demonstrated to us by Dr. Bernard Oliver, vice-president of Hewlett-Packard. Though I was impressed, even awed, I did not fully realize that something revolutionary had come into the world.

30

It is quite impossible for even the most far-sighted prophet to visualize all the effects of a really major technological development. The telephone and the automobile produced quantum jumps in communication and transportation. They gave ordinary men a mastery over space that not even kings and emperors had possessed in the past. They changed not only the patterns of everyday life, but the physical structure of the world—the shapes of our cities, the uses of the land. This all happened in what is historically a moment of time, and the process is still accelerating. Look how the transistor radio swept across the planet within a single generation.

Though they are not yet as important as books, audiovisual aids such as film strips, 16-millimeter projectors, and videotape machines are rapidly penetrating the educational field. Most of these aids are still far too expensive for developing countries, however, and I'm not sure they are really worth the cost of producing them.

Perhaps the most influential device of all is the ordinary television set, whether intended for education or not. I'd be interested to know what impact *Sesame Street* has on the relatively few children of a totally different culture who see it here in Sri Lanka. Still, every TV program has some educational content; the cathode-ray tube is a window on the world—indeed, on many worlds. Often it's a very murky window, with a limited view, but I've slowly come to the conclusion that on balance even bad TV is preferable to no TV.

The power of television lies in its ability to show

31

current events, often as they are happening. But for basic educational purposes, the video recorder is much more valuable. Its pretaped programs can be repeated at any convenient time. Unfortunately, the chaos of competing systems has prevented standardization and cheap mass production.

Videotape machines, however, are far too complicated; they can never be really cheap or long-lived. Video discs, which are just coming on the market, will be much cheaper. Yet I am sure that they, too, represent only a transitional stage. Eventually we will have completely solid-state memory and storage devices, with no moving parts except laser beams or electric fields. After all, the human brain doesn't have any moving parts, and it can hold an enormous amount of information. The electronic memories I'm talking about will be even more compact than the brain—and very cheap. They should be ready soon.

Consider the very brief history of the computer. The first models were clumsy giants filling whole rooms, consuming kilowatts of power, and costing millions of dollars. Today, only 35 years later, far greater storage and processing capacity can be packed into a microchip measuring 1.625 square centimeters. That's miracle number one. Miracle number two is the cost of that chip: not a couple of million dollars, but about \$10.

The change has already begun. Computer-aided instruction is now available in many American colleges and high schools. Consoles with typewriter keyboards allow the student to "talk" to a central computer at any time of day, going through any subject when he feels like doing so, at the rate that suits him. The computer, of course, can talk to hundreds of students simultaneously, giving each the illusion that he is the center of attention. It's infinitely patient, unlike most teachers, and it's never rude or sarcastic. What a boon to slow or handicapped students!

Today's CAI consoles are big, expensive, fixed units, usually wired into the college computer. They could be portable. Already businessmen are traveling the world with attaché case-sized consoles they can plug into the telephone to talk with their office computer thousands of kilometers away. But it is the completely portable and self-contained electronic tutor that will be the next full step beyond today's pocket calculators.

Its prototype is already here, in the nurseries of the affluent West. The first computer toys, many of them looking as if they'd flown off the screen during a showing of *Star Wars,* invaded the shops last Christmas. They play ticktacktoe, challenge you to repeat simple melodies, ask questions, present elementary calculations, and await the answer—making rude noises if you get it wrong. Children love them, when they're able to wrestle them away from their parents. In 1978 they cost $50; now they're half that. Soon they'll be given away as prizes in cereal boxes.

These are toys, but they represent the wave of the future. Much more sophisticated are the pocket electronic translators that first came on the market in 1979, at about the cost of calculators five years earlier. When you type words and phrases into a

33

little alphabetical keyboard, the translation appears on a small screen. You change languages simply by plugging in a different module. The latest models of these machines have even learned to speak. Soon they may become superb language teachers. They could listen to your pronunciation, match it with theirs, and correct you until they are satisfied.

Such devices would be specialized versions of the general-purpose pocket tutor, which will be the student's universal tool by the end of the century. It is hard to think of a single subject that could not be programmed into these devices at all levels of complexity. You'll be able to change subjects or update courses merely by plugging in different memory modules or cassettes, exactly as you can in today's programmable pocket computers.

Where does this leave the human teacher? Well, let me quote this dictum: Any teacher who can be replaced by a machine should be!

During the Middle Ages many scholars regarded printed books with apprehension. They felt that books would destroy their monopoly on knowledge. Worse still, books would permit the unwashed masses to improve their position in society, perhaps even to learn the most cherished secret of all—that no man is better than any other. And they were correct. Those of you who have seen the splendid television series *Roots,* which I hope comes to Sri Lanka someday, will recall that the slaves were strictly forbidden to learn reading and had to pretend that they were illiterate if they had secretly acquired this skill. Societies based on ignorance or repression cannot tolerate general education.

Yet the teaching profession has survived the invention of books. It should welcome the development of the electronic tutor, which will take over the sheer drudgery, the tedious repetition, that are unavoidable in so much basic education. By removing the tedium from the teacher's work and making learning more like play, electronic tutors will paradoxically humanize education. If a teacher feels threatened by them, he's surely in the wrong profession.

We need mass education to drag this world out of the Stone Age, and any technology, any machine, that can help do that is to be welcomed, not feared. The electronic tutor will spread across the planet as swiftly as the transistor radio, with even more momentous consequences. No social or political system, no philosophy, no culture, no religion can withstand a technology whose time has come—however much one may deplore such unfortunate side effects as the blaring tape recorders being carried by pilgrims up the sacred mountain Sri Pada. We must take the good with the bad.

When electronic tutors reach technological maturity around the end of the century, they will be produced not in the millions but in the hundreds of millions and cost no more than today's pocket calculators. Equally important, they will last for years. (No properly designed solid-state device need ever wear out. I'm still using the HP 35 Dr. Oliver gave me in 1970.) So their amortized cost will be negligible; they may even be given away, with users paying only for the progams plugged into them. Even the poorest countries could afford them—especially when the reforms and improved productivity that widespread education will stimulate help those countries to pull themselves out of poverty.

Just where does this leave the schools? Already telecommunication is making these ancient institutions independent of space. *Sunrise Semester* and *University of the Air* can be heard far from their "campuses." The pocket tutor will complete this process, giving the student complete freedom of choice in study time as well as in work location.

We will probably still need schools to teach younger children the social skills and discipline they will need as adults. But remember that educational toys are such fun that their young operators sometimes have to be dragged kicking and screaming away from their self-imposed classes.

At the other end of the spectrum, we'll still need universities for many functions. You can't teach chemistry, physics, or engineering without labs, for obvious reasons. And though we'll see more and more global classes, even at the graduate level, electronics can never completely convey all the nuances of personal interaction with a capable teacher.

There will be myriads of "invisible colleges" operating through the global communications networks. I remarked earlier that any teacher who could be replaced by a machine should be. Perhaps the same verdict should apply to any university, however ivy-covered its walls, if it can be replaced by a global electronic network of computers and satellite links.

But there will also be nexuses where campuses still exist. In the year 2000 many thousands of students and instructors will still meet in person, as they have done ever since the days of Plato's Academy 23 centuries ago.

PART TWO:
HARDWARE

THE BIONIC BRAIN

By G. Harry Stine

Lee's report was due the next day. He'd need Cy to do an entire market study and get the writing finished that soon. Cy, a "cybernetic interface device," had cost far more than a home computer terminal would have, even one hooked into the Library of Congress and the New York Public Library, but the intelligence amplifier was well worth the price. It responded directly to Lee's thoughts, calling up information and taking dictation as quickly as he could think. The work was not interrupted by talking to the computer or by typing on a keyboard. In fact, when Lee had trouble thinking, Cy would often help organize his ideas into coherent paragraphs.

The executive settled back in the interface couch and donned the cap. It took only minor positioning to fit the sensors into place. Then he closed his eyes and pressed the switch built into the armrest.

"Good morning, Lee! What do we work on today?" asked Cy's image, projected directly into Lee's brain and programmed to appear as a synthesis of his favorite college professors and respected business leaders.

"The chief wants a preliminary market analysis for mining the Jovian satellites, emphasis on Ganymede," Lee explained. "What have you got?"

"Everything the Space Industrialization Administration has released to date, including some reports that just came in this morning. May I suggest you add the latest data on mining the cloud tops of Jupiter with scoop ships. Here's a rundown . . ."

After what seemed like hours, Lee stopped the summary so he could tape his report. As he dictated, the computer image occasionally asked whether he wanted to rephrase something. Then Cy projected a series of graphs and color photos from the Jovian system into Lee's visual cortex. Lee selected several, added captions, and ordered, "Okay, put it on the net to the office, and let's call it a day."

He checked the clock. It told him he had been linked to Cy for 28 minutes. A good day's work!

Will electronic computers replace the human brain? Will computer-directed robots make men obsolete? Or will computers take over so completely that human beings are themselves turned into robots? As electronic computer circuits get smaller and more powerful and robots begin to replace human beings in repetitive clerical tasks and manual labor, many people have come to believe that these questions have already been answered and that the human race is coming away with the short end of the stick.

They are wrong. Human beings and computers should not be viewed as antagonists. They're not. The electronic computer is a tool developed by humans that happens to be smaller and faster than the tools it has replaced: pencil and paper.

To date, the computer's power has been applied only to complex calculations or to simple, repetitive chores. That will not always be so. We will eventually build the first intelligence amplifier, a blend of computer and brain, optimizing both. We will link the brain and nervous system directly to the electronic computer, without the cumbersome keyboards, printers, and TV displays we use today. The computer will become not an antagonist but the ultimate extension of our reasoning, memory, and computational ability.

We are closer to building an intelligence amplifier than most people realize. A primitive way to feed information from the human nervous system to a computer has already been worked out, and we may also have the technology to send it from a computer to the brain. It remains only to take these laboratory demonstrations and put them together in the first "interface device."

CRYSTALS VS. COLLOIDS

That is not to say that all the technical details have been resolved. Enormous problems remain, many of them stemming from the great differences between the two kinds of systems we are trying to join. The electronic computer is made up of crystalline, solid-state semiconductors. The atoms in a crystal are arranged in rigid arrays known as lattices. The interatomic forces that hold the lattice together usually make crystalline materials very strong. Most metals, for example, are crystalline. The human nervous system, in contrast, is made of colloids, amorphous, often jellylike materials in which atoms and large molecules are suspended at random. There is no lattice structure in a colloid.

Electronic computers carry information as a flow of electrons through the crystal lattice. Crystalline "brains" are therefore very fast. The human brain codes its information as a relatively slow flow of atoms and molecules through the colloidal mass. Our nerves use two types of data carrier: large molecules, called neurotransmitters, that flow across the synapse, or gap, between nerve cells, and ions, charged atoms that move along the nerve to generate an electrical impulse.

There is one important similarity in the way crystalline and colloidal systems transmit information. Both seem to operate by a binary code. Data in a computer are broken into "bits." An electronic circuit is switched on or off, and all information, no matter how complex, is recorded in this two-unit code. Similarly, a neuron either fires an electrical impulse or does not. There are no in-betweens.

Yet crystalline and colloidal brains process information very differently because of their contrasting structures. Because nerve cells operate by the movement of large, slow atoms and molecules, their reaction times are measured in milliseconds, or thousandths of a second. The fastest nerve cells carry electrical impulses at only 20 meters or so per second. The modern crystalline computer operates in picoseconds, or thousand-billionths of a second. This is a difference of a billion times, or nine orders of magnitude. (This is why a modern computer can operate in a "time-sharing" mode, in which hundreds of humans are working with it at once.)

From a human viewpoint, a computer operates instantaneously. Push the button, and the answer appears, even though the computer has gone through over a million calculations. It is the sheer speed of crystalline systems that appears to bother many human beings, who, by their very nature, act slower, even though they are vastly more complex than a computer.

For a computer, talking with a human being takes a long time. Even with a direct link to the human nervous system, a computer must send its information a billion times slower than it is able to, then wait the equivalent of six years for a reply! If a computer could feel emotions, it would probably be exceedingly bored.

MESHING THE MINDS

Compensating for this speed difference is one of the most important technical problems in creating an intelligence amplifier. Engineers have spent years speeding up crystalline circuitry. Computers operate so quickly that the need to wait while an electron moves a few thousandths of a centimeter is beginning to delay their operations. The state of the art is rapidly approaching the point where the movement of a single electron through the crystal lattice will be enough to transfer a bit of information.

Now, somehow, we must either slow down the computer's crystalline system or speed up the human colloidal system. Fortunately, slowing the crystalline computer presents no problem. Only those circuits that communicate directly with our nervous system must be adapted. After all, this is what the colloidal system does. Our autonomic nervous system doesn't interface with the consciousness and the thinking circuits until we become aware of our heartbeat or other automatic functions.

The system would be more efficient, though, if people could absorb information more quickly. The brain's complexity may make this possible. Unlike computers, which can perform only one operation at a time, the brain compensates for its slow response by splitting up nerve signals and sending them over many channels at once, then recombining them at the receiving end. This technique—electronics engineers call it multiplexing, and they use it in sophisticated stereo and communications equipment—lets the colloidal brain carry out a vast number of operations simultaneously.

Thanks to multiplexing, we may be able to speed our information intake by a factor of ten or more with special training—once we learn enough about human thought processes. We may actually think much faster than simplistic measurements of neuron response suggest. We already know that "psychological time" can be quite different from "physical time."

This whole area of psychological time, human thought processes, and the multichannel nature of our brain is ripe for serious investigation. It is a real pity that psychedelic drugs came along almost simultaneously with one of our culture's periodic swings into Dionysian romanticism. These substances could have become an important tool for this research. They still may, once the furor dies down. The Oriental shaman may have learned to control psychological time ages ago. If we cannot gather good, solid data in the area from highs, trips, and mysticism, perhaps we can be led to it by the distrusted computer and intelligence amplifier.

All this assumes, of course, that we can actually link the colloidal and crystalline brains, and there is a good chance that we can. Since both systems encode information as electrical charges, they produce electromagnetic fields that we can detect, manipulate, and perhaps translate from the brain's "language" to that of the computer, and back. Scientists have already made a strong start.

READING THE BRAIN

Last year the Defense Advanced Research Projects Agency (DARPA) reported to Congress "significant progress" in an area called biocybernetics. DARPA researchers have managed to extract useful information from the brain's electrical activity. The electroencephalogram (EEG) is measured by electrodes placed on the scalp or inserted into the brain itself. Although it has been known that the EEG varies with mood—we generate alpha rhythms when relaxed, for example—trying to relate the mysterious squiggles of an EEG to specific thoughts and motor processes has seemed futile.

Recently, though, DARPA researchers have used computers to identify the EEG signals that distinguish thinking, or cognitive, processes from motor responses, such as signals to the muscles. By measuring the EEG signals for motor responses and those for cognitive load, they have been able to assess spare brain capacity from moment to moment.

DARPA also claims that computers can identify EEG waves associated with decision making and action. If the EEG's decision-making component ends before the action component, the researchers say, the decision was probably correct. But when the decision-making signal continues after the action component ends, the probability of error is very high. A computer can now tell its operator, "Excuse me, Joe. I think you made a mistake there."

Modern electronics will probably tell us much more about how our elusive neural signals reflect our thoughts and feelings. Scientists can now investigate EEG frequencies of up to several million hertz. This is an enormous increase in sensitivity over the 100-hertz range of the classic EEG chart recorders.

Despite these refinements, it seems unlikely that the EEG will ever be able to operate a human/computer intelligence amplifier. The voltage differences measured by the EEG's widely separated electrodes cannot be traced to specific brain locations—a must for any useful link.

Physiologists have always yearned for a technique that would directly monitor localized brain activity from a distance so as to avoid interfering with the brain's normal function. Today they have it.

Electrical activity always produces a magnetic field around it. The brain's currents are no exception. Twenty years or so ago researchers went looking for magnetic fields created by biological processes. In 1963 Gerhart Baule and Richard McFee reported in the *American Heart Journal* that they had detected the biomagnetic field of the beating heart. But the field was only one-billionth the strength of the earth's magnetic field. To detect it required extensive shielding; even then, sensing coils with 2 million turns of hair-fine wire could barely pick it up.

Today the superconducting quantum interference device (SQUID), which uses superconducting niobium coils cooled in liquid helium, is more than 1,000 times more sensitive to magnetic fields. By coupling SQUIDs with other modern electronic techniques, Dr. Lloyd Kaufman and his colleagues at New York University have eliminated environmental magnetic noise almost completely.

A SQUID positioned a centimeter from the scalp can produce a magnetoencephalogram (MEG) far more sensitive than an EEG. Kaufman and his fellow researchers can now locate neural activities in the brain within several millimeters. So far, Kaufman has mapped the response of the visual cortex to simple stimuli and has located the brain's reactions to electrical currents applied to the fingers.

Eventually the MEG responses of the entire brain and the spinal cord will be mapped. At that point, computers may be able to decipher our MEGs and read our minds—if we let them.

THE FLANAGAN AFFAIR

But how about the other way around? How can computers talk directly to the human nervous system? On July 24, 1962, I had my own nerves linked to an electronic circuit that fed audio signals directly into my brain, without loudspeakers and without any electrical connection.

A teenage gadgeteer named G. Patrick Flanagan, of Bellaire, Texas, had stumbled onto creating what he called a neurophone. Because no one had any idea how the device worked, it seemed very complex, but, technically it was very simple, nothing more than a 35-kilohertz oscillator amplitude modulated by a hi-fi amplifier. The amplifier fed the combined signals from the oscillator and the amplifier through a transformer that produced an output with very high voltage and very low amperage.

An ordinary TV antenna wire carried the signal to two insulated pads that Flanagan had taken from a muscle-relaxer device. The 7.5-centimeter pads were basically a sandwich of metal mesh connected to the TV lead and insulated by two rubber disks. If you put one pad on your spine and the other on the sole of one of your feet, you could hear perfect hi-fi in your head the moment contact was made.

I investigated the Flanagan neurophone as a possible new project for a small industrial firm. In three years of complex experiments researchers concluded that bone and skin conduction had nothing to do with the transmission of audio information to the nervous system.

Dr. Wayne Batteau, then at Tufts University, proved later that the neurophone was directly activating the human nervous system and that the audio information was not being picked up and transmitted to the brain via the auditory nerve. In fact, Dr. Batteau reportedly restored hearing to a nerve-deaf patient.

Somehow the Flanagan neurophone seemed to couple electronic circuitry directly to the human nervous system. The device could apparently send audio information along any nerve path to the brain, which recognized the signal as audio data and switched it to the appropriate area of the cortex.

Unfortunately, my own research with the neurophone ended abruptly when the company I worked for decided the project did not fit in with its product mix. Shortly thereafter Dr. Batteau died of a massive myocardial infarction while scuba diving with dolphins in Hawaii, and Flanagan became involved with Oriental mysticism and developed into a leading exponent of "pyramid power."

Unbelievable as the Flanagan neurophone may sound, I can assure you that it was no hoax. Many responsible people experienced its effects. The experiments were conducted under the most controlled conditions we could arrange.

In 1962 the neurophone, far ahead of its time, was considered only as a new type of hearing aid. Although it has remained unused for more than a decade, I hope that interest in it will be renewed and that research will resume. Technology may now have caught up to the point where the neurophone could be used as the basis for human-to-computer interface.

TO RUN THE WORLD

Of course, it is still far too early to say whether the intelligence amplifier will even remotely resemble Cy. But one thing is certain: Advances in our knowledge and technical skills are bringing us closer to a working, fully functional interface device.

The intelligence amplifier will combine our creative, self-aware, multichannel, and many-circuited nervous system with the high-speed computation of the electronic computer. The crystalline computer will become an extension of our own minds, a new tool to expand intelligence.

Viewed in perspective, the intelligence amplifier is only a logical step in the evolution of computer technology. Computers and robot machines have taken over much of manual labor and painstaking computational work. Yet we have only begun to explore the computer's ability to reduce our mental work load. So much of our time and resources, especially during our education, are still devoted to memorizing an enormous body of information and ideas that forms the basic framework on which all later knowledge is built. Why shouldn't we use computers to help us.

Will the crystalline computer in an intelligence amplifier take over and rule the colloidal, human portion? Only if we humans design it to do so. The computer is a tool . . . and, yes, tools occasionally get out of hand. The hammer can bang your thumb if you aren't careful. Fire can burn you. But because tools are not always safe is no reason not to have and use them.

Perhaps one last question must be asked: Why try to build the ultimate computer, the intelligence amplifier, in the first place? Why not continue to rely on ordinary human intelligence?

To help me keep some perspective about the world, I've put a motto in Latin above my desk: *Nescis, mi fili, quantilla sapientia regitur mundus.* Rather loosely translated, this tells me, "You'll never know, my son, with what little real knowledge the world is run."

To run the world better, with more real knowledge, we need all the help we can get. That is the real purpose of the intelligence amplifier.

TELEPRESENCE

By Marvin Minsky

You don a comfortable jacket lined with sensors and musclelike motors. Each motion of your arm, hand, and fingers is reproduced at another place by mobile, mechanical hands. Light, dextrous, and strong, these hands have their own sensors through which you see and feel what is happening. Using this instrument, you can "work" in another room, in another city, in another country, or on another planet. Your remote presence possesses the strength of a giant or the delicacy of a surgeon. Heat or pain is translated into informative but tolerable sensation. Your dangerous job becomes safe and pleasant.

The crude robotic machines of today can do little of this. By building new kinds of versatile, remote-controlled mechanical hands, however, we might solve critical problems of energy, health, productivity, and environmental quality, and we would create new industries. It might take 10 to 20 years and might cost $1 billion—less than the cost of a single urban tunnel or nuclear-power reactor or the development of a new model of automobile.

To convey the idea of these remote-control tools, scientists often use the words *teleoperators* or *telefactors.* I prefer to call them *telepresences,* a name suggested by my futurist friend Pat Gunkel. *Telepresence* emphasizes the importance of high-quality sensory feedback and suggests future instruments that will feel and work so much like our own hands that we won't notice any significant difference.

Telepresence is not science fiction. We could have a remote-controlled economy by the twenty-first century if we start planning right now. The technical scope of such a project would be no greater than that of designing a new military aircraft.

A genuine telepresence system requires new ways to sense the various motions of a person's hands. This means new motors, sensors, and lightweight actuators. Prototypes will be complex, but as designs mature, much of that complexity will move from hardware to easily copied computer software. The first ten years of telepresence research will see the development of basic instruments: geometry, mechanics, sensors, effectors, and control theory and its human interface. During the second decade we will work to make the instruments rugged, reliable, and natural.

Three Mile Island really needed telepresence. I am appalled by the nuclear industry's inability to deal with the unexpected. We all saw the absurd inflexibility of present-day technology in handling the damage and making repairs to that reactor. Technicians are still waiting to conduct a thorough inspection of the damaged plant—and to absorb a year's allowable dose of radiation in just a few minutes. The cost of repair and the energy losses will be $1 billion; telepresence might have cut this expense to a few million dollars.

The big problem today is that nuclear plants are not designed for telepresence. Why? The technology is still too primitive. Furthermore, the plants aren't even designed to accommodate the installation of advanced telepresence when it becomes available. A vicious circle!

Perhaps you have seen the current style of remote-control arms used at nuclear facilities. They are little better than pliers, unable to do many things you can do with your own hands. Anyone can buy a simple remote manipulator off the shelf. It usually consists of an input unit for the operator to control and of an output device that does the work. Typically, the input is a handle attached to a jointed armlike linkage. When you squeeze the handle, a gripper closes at the output. But no such device demonstrates true telepresence. The remote gripper may well imitate the motion of your hand, but the remote arm does not follow your arm's curve, and so you cannot always reach around obstacles. The dynamics are unnatural, and the designs skimp on many shoulder, elbow, and wrist motions. The hands have unnatural wrists. The conventional grippers can pinch or clamp but can't twist, shear, roll, or bend. They can't use ordinary scissors. Instead, someone has to remove the hand and replace it with a special tool for that particular task.

If people had a bit more engineering courage and tried to make these hands more like human hands, modeled on the physiology of the palm and fingers, we could make nuclear-reactor plants and other hazardous facilities much safer.

My first vision of a remote-controlled economy came from Robert A. Heinlein's prophetic 1948 novel, *Waldo*. I suppose Heinlein had heard about myasthenia gravis, a disease causing profound muscle weakness. His hero, Waldo, a wealthy young man, was afflicted with it. So Waldo constructed a satellite and invented telepresence devices; he could lie there in zero gravity and operate his inventions effortlessly. Waldo created dozens of mechanical hands, some merely monkey fists in size, some micrometers in span; he rigged others so huge that each "hand" spread six meters from little fingertip to thumbtip. The hands imitated everything he did; he spent all his time out in space operating factories on Earth. Thirty years after he wrote *Waldo,* Heinlein had many suggestions for this article.

Developing telepresences will involve hard scientific and engineering problems, but I believe we should go ahead. Present devices are so clumsy, they are used only when nothing else works. Once improved, however, telepresence will bring us:

•*Safe and efficient nuclear-power generation, waste processing, and land and sea mining*. Last year's Gulf of Mexico undersea oil blowout is the kind of accident that I'm convinced telepresence technology could have helped to mitigate.

•*Advances in fabrication, assembly, inspection, and maintenance systems*. With telepresences one can as easily work from a thousand miles away as from a few feet. Manual labor could easily be done without leaving your home. People could form "work clubs." One region of the world could export the specialized skills it has. Anywhere. A laborer in Botswana or India could market his or her abilities in Japan or Antarctica.

•*The elimination of many chemical and physical health hazards and creation of new medical and surgical techniques*. If we miniaturize telepresence for use in microsurgery, for example, surgeons could repair or replace many little blood vessels in the brain. Other organs beyond the reach of scalpel and forceps could similarly be repaired or substituted.

•*A reduction of transportation costs and of energy and commuting time, enabling one person to do different jobs in different places*. Mass transportation could be replaced by ubiquitous taxis, remotely controlled by teleoperation. Telepresence devices could fix sewers, electrical conduits, and water mains from within. Teleoperation will do away with all hazardous and unpleasant tasks.

•*The construction and operation of low-cost space stations*. Telepresence might prove invaluable for solar-power satellite construction—for amassing materials in space and supplies for the human work force. Telepresence would be able to assemble various orbital structures. There are many places here on Earth more dangerous to humans than outer space is. Mines, for example. In a remote-controlled mining operation, there are no people to be hurt. A fire in a mineshaft or a collapse would elicit no more response than: "Well, it is very sad. We've lost six robots."

Remote-controlled mining can exploit buried resources efficiently and humanely. No one is exposed to the dangers of explosions or of breathing in coal dust. No more blacklung disease. We will mine the meter-thin deposits of anthracite coal now lodged in formations we cannot reach. Underground combustion and gasification schemes, which are presently unfeasible because they cannot be controlled, may then be feasible.

The biggest challenge to developing telepresence is achieving that sense of "being there." Can telepresence be a true substitute for the real thing? Will we be able to couple our artificial devices naturally and comfortably to work together with the sensory mechanisms of human organisms?

When any job becomes too large, small, heavy, or light for human hands, it becomes difficult to distinguish the inertia and elasticity of the instrument from what it's working on. Telepresence will be able to adjust and compensate for such problems, thus making the job easier. For instance, a remote "miner" could dig a narrow seam without himself having to stoop or crawl. Machines will incorporate new theories of human sensory pattern perception and feedback control to "reflect" accurately to the user the modified remote sensations.

We have talked of mining, but no matter how much coal we mine, we are, like it or not, becoming dependent on nuclear power. Even if it were to be banned in the United States, we cannot prevent its proliferation abroad. The nuclear designers try to anticipate and avoid all modes of failure. But all reactors have the potential to break down: High temperatures weaken structural materials; generators apply high pressures to those weakened structures; and radiation damage makes inspection difficult, while aggravating structural damage and corrosion.

These problems compel designers to choose between two extremes. One is to build each part with monumental toughness—to minimize human exposure—and hope this system never fails; this is today's designers' favorite approach. But in the end breakdown and failure occur anyway, requiring man's intervention. Even a minimal failure shuts down a reactor for months.

I think the better extreme is to build modular systems that permit periodic inspection, maintenance, and repair. Telepresence would prevent crises before they could arise.

If no one were in the buildings, no one would be exposed to radiation. Then we could all stop quarreling about "tolerable" and "threshold" doses. If nothing enters or leaves the reactor except by way of telepresence machines, no one can steal anything. Computers—or skeptical people—can monitor for unusual activities over viewing channels. This allows few opportunities for sabotage, and it makes it easier to combine power generation, fuel processing, and waste management.

We can employ telepresence in any environment alien to humans. Most of the earth, for example, is ocean; "moonwalks" on the ocean floor at two miles' depth are technically more difficult to execute than moonwalks on the moon or Mars. Remotely operated seafloor "construction crews" could bypass the prohibitive hazards of manned exploration, avoiding the risk of weather-troubled ships and treacherous towers in mining on the continental shelf. The U.S. Office of Naval Research has some remote-controlled deep-sea exploration projects, and eventually such systems will explore for and extract deep-sea petroleum and minerals. Eventually entire undersea industrial plants could be so controlled from the surface.

There are already some undersea manipulators. The Alvin submersible, at Woods Hole, Massachusetts, is wonderful, but its manipulator is used mainly for picking up samples. You couldn't tie a knot with it. I'd like to see one that can do anything fingers can do.

In space the amazing success of *Vikings 1* and *2* shows how much can be done with remote control —even with day-long transmission delays. Yet the *Viking* spacecraft had pathetic limitations. There was no way to reconfigure the equipment to make use of what was learned; a week of breathless planning was required just to get *Viking 2* to turn a stone over.

I think the best way to explore the planets is to

have people in orbiting spacecraft to operate telepresences that maneuver on the surface. A Mars Rover with good telepresence manipulators can make extensive excavations, then reconfigure scientific equipment to exploit what has been discovered.

Think how much more we could have learned with a permanent vehicle on the moon. The Earth-Moon speed-of-light delay is short enough for slow but productive remote control. With a lunar telepresence vehicle making short traverses of one kilometer per day, we could have surveyed a substantial area of the lunar surface in the ten years that have slipped by since we landed there.

Among the most exciting prospects for solving our energy woes is to build a ring of solar-power satellites in orbit around Earth. Safe, free solar energy could then be collected and beamed back to receivers located near our cities. The main problem is the cost: We must put sufficiently large structures in space to gather enough sunlight, since each station requires thousands of acres of reflectors and collectors. And then there's the cost of sending people into space to build them.

Telepresence could save billions of dollars by employing remote-controlled hands stationed in orbit and controlled by technicians on Earth and on the moon. Most satellite construction could be done by people working in their own homes and offices.

To circumvent the cost of lifting satellite payloads against Earth's powerful gravity, scientists have been devising ways to manufacture and launch materials directly from the moon or from the asteroids. Building such lunar facilities, however, would be impossibly expensive if carried out entirely by men in space suits on the moon. Instead, why not use cheap, Earth-based labor via telepresence to build moon factories? Imagine having to go no farther than your study to operate a crane on Mare Imbrium. We need only send telepresence machines on inexpensive one-way trips.

The scenario includes sending 20 real men and women to the moon. It's not very difficult. Saturn 5's have the potential to send up a crew of scientists and engineers with many months' supplies. (If only they didn't have to bring them back!) So we ferry up a return vehicle for use in emergencies. Then we send up more permanent housing and, finally, the superflexible telepresence equipment needed to construct the first lunar installations. The people are there to supervise the work and to fix the equipment when necessary.

One major obstacle to all this has been NASA's legislated inability to deal with such far-reaching concepts as telepresence. The U.S. space program is entirely mission-oriented. NASA never gets appropriations to make better manipulators or navigational devices, as things in themselves. Even so, scientists at Ames Research Center, in California, managed to develop a startlingly nice telepresence, a remote-controlled space suit. It looks like a real space suit; you put your arm into the master suit and the slave suit moves just like your arm. It's an extremely good arm, a perfect imitation. Your arm feels natural in it. But it doesn't have any hand.

The space shuttle, too, has an arm. It is very long, and it takes about half a minute to complete any motion. But there isn't any reason to hurry. In zero gravity nothing weighs anything; so one can use a 100-pound, long, slender pipe to move a 10-ton load very slowly. (There's a simulator for this at Marshall Space Flight Center, in Alabama. It is a model of the fuel tank of the space shuttle—a helium balloon about as big as a house. Sitting in its hangar, it weighs nothing. When you press on it, no movement occurs for about 30 seconds, but then it begins to move. If you grab on to it, it pulls you right up, too, and because it has no weight, you can lift it with your hand, but it has a mass of half a ton!)

While the shuttle arms are merely glorified construction cranes, they are the beginnings of giant teleoperators. At the other end of the size spectrum, biologists have long used micro-manipulators, tiny teleoperators. But none of them have any sensors. If we were to miniaturize telepresences for surgery, we could develop touch-reflecting microhands on slender probes that reach through the vessels' narrowest passages. Further in the future a surgeon could direct a semi-intelligent procedure, including several simultaneous micro-telepresences, to make smaller repairs swiftly.

The first crude remote-controlled mechanical hands were built around 1947 at Argonne National Laboratories, in Illinois, for handling dangerous chemicals. In 1954 the late Ray Goertz, a scientist at Argonne, developed hands in which electric motors "reflected" some of the forces back so that the operator could feel something of what was happening, at least resistance and pressure, if not textures. Paradoxically, the very first telepresences could relay sense of feel better than later electric models could, because they used rigidly linked cables and pulleys. Later electric motors were stronger and could work at greater distances, but they lost that sense of feel one got through the cables. More advanced models measured forces at the output and used additional motors to reflect those forces back to the user's hands. When the remote claw hit something, the input became harder to push. This helped, but the force reflection was still inadequate for performing delicate work.

Early pioneers like Goertz had the fantasy of building better robots of various kinds, and then people got interested in my field, artificial intelligence—getting robots to do smart things. And we did get them to do simple mechanical things, some factory work, like assembling a motor. But they were always handicapped by those terrible claw hands.

To create true telepresences, we must supply more natural sensory channels—touch, pressure, textures, vibration. We must learn which sensory defects are most tolerable. In 1958 Ralph Mosher, an engineer working for General Electric, developed a telepresence—called Handiman—that had good dexterity and compensation. It had only two fingers, but those fingers each had three joints so that they could wrap around any object. Handiman could lift hundreds of pounds; it transformed you into a superperson. But it was never put to any practical use. Mosher subsequently made a simpler version that permitted him to sit in his chair and pick up refrigerators.

Another big manipulator was designed and built in the late Fifties as part of a project to build nuclear airplanes. But Congress finally decided the plane, designed to stay aloft for a year without landing, wouldn't be safe.

Although basically little work has been done since the Fifties, there now exist a few more versatile experimental manipulators. Electrotechnical Laboratory, in Tokyo, has made a three-fingered, 12-jointed hand that can roll a baton. But that's about all it can do. A group at Stanford University invented a long, snakelike tentacle that can wrap around objects. I once built a 14-joint, multi-elbowed arm that can easily reach around things in its way. But no project has the resources to perfect any such ideas.

I think we should make telepresences that compare well with the human hand: a five-fingered device capable of imitating natural motions. It should be mobile. We might then adapt designs and concepts from the arm to make legs, yielding a system able to work wherever people can, not only on carefully prepared floors.

To control such an instrument, we will want a light, well-articulated sleeve that includes effectors to reflect the sensations. This will require advanced materials and new muscle-imitating devices; for visual feedback, we'd need slender fiber-optic probes articulated to emulate the operator's head-and-eye motions. We probably would want to have an eye of some sort on the fingers.

A Philco engineer named Steve Moulton made a nice telepresence eye. He mounted a TV camera atop a building and wore a helmet so that when he moved his head, the camera eye on top of the building moved, and so did a viewing screen attached to the helmet.

Wearing this helmet, you have the feeling of being on top of the building and looking around Philadelphia. If you "lean over," it's kind of creepy. But the most sensational thing Moulton did was to put a two-to-one ratio on the neck so that when you turn your head 30 degrees, the mounted eye turns 60 degrees; you feel as if you had a rubber neck, as if you could turn your "head" completely around!

Why did telepresence stop evolving 20 years ago? One reason is that research funds declined while costs escalated; by 1960, no laboratory could afford to take another step. But a more fundamental reason for this stagnation is that engineers are far too clever at solving immediate problems. This has led to endless repetition of the same scenario: An application needs a better manipulator, for example, to join two pipes together; an existing mechanical hand would help, but only if it had another joint in one of its fingers. One must add another control channel, design a new sensor and input feedback device, modify a microcomputer program, retrain the operators, and reengineer the older tools. All this puts enormous strains on a company's budget. Ultimately pipe fittings are redesigned so that the old, clumsy hand can manage.

A good production engineer can solve almost any specific problem by using a special jig, fabricating a new part, developing a special tool for the hand, or replacing the hand with a special tool. Each problem gets solved, to be sure, but the overall technology becomes antiquated and goes unnoticed until an accident such as the one at Three Mile Island or the Mexican oil spill occurs and we find out there is no way at all to turn a valve from afar, or to replace it.

Several major companies have been involved in telepresence research from time to time—AMF, Hughes, General Mills, IBM, and others—though none of them have reached "critical mass." Many smaller firms possess more precious skills— Unimation, Central Research Laboratories, Programmed and Remote, and others. The Defense Advanced Research Projects Agency, working with the army, once supported work on powered armored suits, like those in Heinlein's *Starship Troopers,* but the work was abandoned. University workers have had many good ideas, designs, and prototypes, but they could never afford to engineer complete systems. There are important projects at Carnegie, MIT, Jet Propulsion Laboratory, Stanford, and other university labs.

Part of the problem has been that telepresence has never been anybody's baby. Such a project demands centralization. It requires imaginative specialists in sensors, effectors, control theory, artificial intelligence, software, engineering, psychology, and first-class facilities for mechanical and electrical engineering and materials science. It will need strong resources for interactive computation and for real-time physical simulation.

It would be difficult to assemble such an organization in today's peacetime atmosphere. What we need is a modern league of working centers connected by a computer network. Such a network would contain a central data bank somewhere in the United States, combining administrative and engineering resources that need centralization. Perhaps there could be computer centers close to universities or industrial locations, working together through communications networks such as ARPANET (the computer conferencing network of the artificial intelligence community).

I can't imagine anyone doubting that telepresence is possible. It's a matter of solving many problems that are hard, but not impossible. In the mechanical area the same things have been done over and over; engineers spend their time arguing about what kind of wrist is better. Some think the wrist should go round and round, spinning forever, especially on a robot that's using a screwdriver. That's all right for the industrial robots, but there isn't much point in that for a telepresence, because you can't spin your own wrist around to control it.

Some research has been done in the psychology of spatial perception, in terms of feedback controls and the interrelationship between electronics and the human nervous system. There is a device, for example, that can translate print into "feel," developed by J.C. Bliss and J.G. Linvill, at Stanford, which enables the blind to read conventional printed matter. It's a gadget that fits conveniently on your fingertip and has a lot of miniature photocells to sense light and a lot of little vibrators that allow the finger to sense remotely the fine shape of the letters. In my own laboratory, graduate student Danny Hillis recently fabricated a thin, skinlike material that can "feel" and transmit small tactile surface features.

Someone could develop similar devices for telepresence—vibrating patterns, for example, that would convey the sensations of "hot" or "cold." However, very little is known about tactile sensations. It seems quite ironic to me that we already have a device that can translate print into feel, but that we have nothing that can translate *feel* into feel.

Eventually telepresence will improve and save old jobs and create new ones. Later, as we learn more about robotics, many human telepresence operators will be able to turn their tasks over to the robots and become "supervisors." In the long run, since each step toward telepresence is a step toward robots, telepresence sensors and output devices could be controlled by computers rather than by people. This becomes inevitable as we learn more about artificial intelligence.

Computers equipped with artificial hands and eyes have actually grasped and moved objects in accord with verbal commands. A complicated precision bearing was so assembled at MIT; an entire pump was put together at Stanford; a toy automobile was constructed in Edinburgh, Scotland. Similar work has been done at SRI International. These laboratory programs are too undependable for practical use because, although we can do many things with computers, we cannot get them to do many things any child can do. Someday our machines could do all our work for us, but that is a long way off, and it would need another whole article to begin to explain the problems.

If teleoperator technology promises wealth and freedom beyond dreams, is there a dark side? People who issue manifestos should think about such matters. The solution may be to grant those who want to live in the "old ways" their chance, while those who want new gifts should also have theirs. I think the gifts promise better, richer, and longer lives. Might telepresence, though, have a special tendency to make workers feel alienated? Perhaps, yes, even with superb technology. Many jobs will become intensely more interesting and more creative; many worlds will be expanded.

If each step toward telepresence were also a step toward the economic pain and psychic grief of unemployment, one might consider working against it. Yet a generation of reforms is already eliminating many of the unsafe jobs that telepresence could preserve. Telepresence offers a freer market of men's and women's skills, rendering each worker less vulnerable to the moods and fortunes of one employer.

Finally, in a strange sense, the question of

"technological unemployment" may become moot: Many young people today consider it demeaning to be bound to any single employer, occupation, or even culture. Perhaps many of us sense—at least on some level—that little of what we do really has to be done. Our attitudes about work, about changing the quality of it, depend as much on our own dispositions and our alternatives as on the jobs themselves. In effect, most of us already feel technologically unemployed.

Postscript: from the *London Telegraph* Foreign Service

PARIS—The French government has authorized Electricité de France, the French central electrical generating board, to go ahead with loading two new nuclear-power stations with enriched uranium fuel. The approval came despite the detection [in 1978] of cracks several millimeters wide in the tube end-plates of the steam generators and in the tubes connecting them to the reactors.

The existence of the cracks first was disclosed . . . by nuclear engineers who pointed out that once the reactors go into operation, repairs would be impossible for lack of appropriate robotic equipment. François Kosciusko-Morizet, government director of industrial quality and security, countered that the defects were very carefully examined and were found to be superficial. In the very worst of hypotheses, the cracks would not give trouble for five or six years, by which time repairs would be easy, he said, *since from 1981 on, France* would have robots, able to repair such defects automatically. [Italics mine—M.M.]

Since France is walking the tightrope of possible power shortages and Electricité de France has warned that possible power cutbacks may be necessary . . . the government has decided that the economic risk of holding up the electricity generation timetable by repairing the reactor cracks now is greater than the danger of serious accidents later.

My work in artificial intelligance is carried on in a world as much fiction as science. This essay used specific suggestions from Isaac Asimov and Robert A. Heinlein as well as from Carl Sagan, Brian O'Leary, Edward Purcell, and many others.

For more technical details I recommend *Remotely Manned Systems,* edited by Ewald Heer (Caltech, 1973), and *Human Factors Applications in Teleoperator Design and Operations,* by E.G. Johnsen and W.R. Corliss (Wiley, 1971). For a discussion of intelligent machines, see M. Minsky, "Computer Science and the Representation of Knowledge" in *The Computer Age: A Twentieth Century View,* edited by M. Dertouzos and J. Moses (MIT, 1979).

THE SIMULATED SOCIETY

By Hal Hellman

It was the first time Dr. Caldwell had ever tried a lens-implant operation, and he was nervous. Eye operations are always difficult, but one slip here could leave the patient permanently blind.

Looking through a microscope, he guided the surgical instruments with knobs and handles. It still felt artificial to him; he had always felt more comfortable when he could hold the tools in his hands. Yet he knew that when doing microsurgery, particularly on the eye, even the steadiest of hands shake too much to operate directly on the delicate tissues.

Peering intently through the stereomicroscope, he was now ready to make the final, major incision. Dr. Caldwell took a deep breath and began to cut. A moment later he gasped, "Oh, no!" and watched helplessly as the laser scalpel bit too deeply into the patient's cornea.

He looked up at Dr. Don Farmer, professor of microsurgery at the University Hospital, who laughed. "I thought you were working a little too quickly," Dr. Farmer said. "Take a five-minute break and try it again."

Again? Yes. Happily, Caldwell was not operating on a live patient, which is what resident surgeons usually must do today. He was working on a computer-operated simulator, a device that was able to mimic the entire operation and respond to his every move with appropriate reactions.

While this surgical trainer is well beyond our present capability, the simulation process is already at work in numerous applications, and new ones are constantly being developed, many of which are touching your life already. As vehicles and equipment become more massive, complex, and expensive, training people to operate them becomes a costly, sometimes dangerous, undertaking. Operators of jumbo jets and oil tankers must learn how to avoid accidents, without ever having the luxury of learning from experience.

A spokesman for the Redifon Corporation, a major producer of simulators, suggests, "With human error estimated to account for the loss of three thousand ships in the last ten years, the need for improved maritime training is obvious."

FROM CAR SEAT TO COCKPIT

Already there are simulators to teach people how to operate every type of moving vehicle—including trucks, tanks, surface vessels, submarines, railway and subway trains, and, if we include the driver-training systems coming into wider use, automobiles. Fifty thousand lives are lost every year in auto accidents. This enormous toll has convinced many people that better training of drivers is a must, requiring perhaps even periodic recertification, as is done with aircraft pilots today.

In present driver-training simulators, the student gets some sense of driving in that the gas pedal, shift lever, brake, steering wheel, turn signal lever, and other functional parts of a car's "cockpit" are simulated, and the student's handling of them is carefully monitored. While "driving," the students observe a film that unravels an actual route before them. The driver is thus constrained to follow the route that has been photographed, and so he gets little sense that he is actually in control of the drive. Since his seat is motionless, he must operate without the proprioceptive sensations that would accompany "real" acceleration, turning, and stopping.

Still, these simulators are becoming indispensable in teaching people how to drive. It's not so much that good "defensive" driving techniques can be taught better in simulators; it's simply that there is no other way to teach them. Reading about them and seeing pictures just don't do it. One must develop the proper reflex actions as well as appropriate reasoned responses in emergencies, and these come only from experience and practice. Anyone who has had a lot of practice with emergencies in a real car is clearly doing something wrong.

The most expensive, most advanced simulators are being used to prepare people to operate the most expensive, most advanced technology.

Although I had never before piloted an airplane, recently I took the controls of an Eastern Airlines 727 jet in Miami and flew it to Washington, D.C. It was a very pleasant trip—except for some rather annoying turbulence, a flameout in one engine, a leaky fuel line, and a sticky rudder. As we readied to land at National Airport—a beautiful sight from the air—I said to the pilot, "I want a hard landing." He grinned, then told the flight engineer to hold on.

We came roaring in. Though I could see the runway coming up to meet us at obviously too steep an angle. I was totally unprepared for what happened next. We *really* hit that runway—there is no other way to describe it—bam! We bounced up some ten meters, then hit again. Bam! And a third time, then finally held the ground. But because of our high speed we were nearing the end of the runway. My "copilot" was applying his brakes hard and told me to hold on. I was pitched forward in my seat. We came to a screeching halt, just a few meters from disaster.

My heart was pounding. Finally, I breathed easier, thankful that we had never left Miami. I was in a flight simulator, a cabin-only aircraft attached to hydraulic actuators, digital computers, and a couple of televisionlike scene generators. The illusion was so complete that when I stepped out of the simulator room and saw that I was in a building with desks, secretaries, and drinking fountains, it took me some moments to shake the disorientation.

* * *

BETTER THAN REAL

Bill de Decker, who trains pilots in simulators for Flight Safety International, maintains that designing a simulator is actually harder than designing the plane itself, "Simulations," he says, "is both art and science. Not only must an advanced aircraft simulator be able to do everything a real plane can do—including simulation of hundreds of problems, emergencies, and malfunctions—but it must all be done by invention." The pressure gauges must work, but a malfunction is provided by electrical command from the computer, not from a real lowering of pressure.

If the instructor decides it's time for a flameout in one engine, the unbalanced forces that would act on the real aircraft must be felt in the controls of the simulator. If the pilot reacts too quickly or slowly, then whatever would happen in the aircraft must happen in the simulator.

The simulator I rode mimicked reality by moving my seat in a variety of ways to correspond to the operation of the controls. These are broken down into three straightline movements (up/down, left/right, and front/back) and three rotational ones (roll, pitch, and yaw, or direction). A simulator with the capacity to mimic all these motions is called a six-degree-of-freedom device.

Hit the brakes in a real plane and it will slow down because of friction forces between the wheels and the runway, and you, because of your inertial motion, will be thrown forward. When you hit the brakes in the simulator, the whole cabin moves backward about half a meter. In the seat, you feel as if you are moving forward. This slight movement is enough to make you think you are being thrown forward; your body doesn't have to go through the whole motion. This "hint that fools" is called an initial cue; it starts your body and mind believing that deceleration is actually occurring.

If the simulator ended its motion after that half meter or so, you would just as abruptly stop believing in the deceleration. Therefore, a sustained motion cue must be added. Now, if we are truly decelerating, something else happens, namely, we are pitched forward. This, too, can be simulated, by having the whole cabin tilt forward. To some extent the cabin would do this in reality, and from the outside you would actually see the aircraft's nose angle down a few degrees (as a car does) during the stop. This position is held for the whole time and distance during which the deceleration is taking place, which can be hundreds of meters for a large aircraft like the 727.

In the meantime the cabin is "sneaking back" to its neutral, or ready, position in a forward direction. Because of everything else that is going on, we don't realize that this is happening. By the time the simulated stop is complete, the cabin has moved back to its original position, and it now pitches upward slightly, to make us think we have fallen backward into the seat.

The tremendous bounce we took while "landing" the Eastern Airlines simulator in Miami was produced in essentially the same way—although faster and more forcefully. At the same time the visual image "out the window" is correlated with the pilot's actions and the movements of the "airplane," making the whole illusion very compelling and lifelike.

SPACECRAFT

While astronauts now tend to be engineers and scientists, the first ones came from the exalted ranks of aircraft test pilots. These men generally flew by the seat of their pants and learned from experience as much about flying as there was to know. So it is not surprising that the first astronauts had strong doubts about the value of simulators. Often, for example, Gus Grissom would hang a lemon on the first Apollo simulator before "flying" it, just to be sure everyone knew how this test pilot felt about playing Let's Pretend. However, Grissom later changed his mind.

Gerald Griffin, flight director of the Apollo 13 mission, has said that "the greatest untold story of the mission is that of the spacecraft simulators." Not only did these devices turn out to be indispensable in the training of astronauts, but they made possible one of history's most dramatic rescues. That occurred when three astronauts were trapped in a moonbound craft that had been crippled in an explosion. The *New York Times* reported: "The countless improvisations that nursed the crippled spacecraft along were in large measure the product of an extraordinarily elaborate assembly of simulators at the Manned Spacecraft Center, in Houston, and elsewhere. Every makeshift procedure carried out in space was first tried out on

Earth and rejected if the simulators showed it to be dangerous or impractical.''

Space-shuttle operators are also being trained in simulators that re-create the extraordinary environment of space in extraordinary ways. Huge vacuum chambers simulate cold and low-pressure conditions, while large water chambers that mimic weightlessness are being used to determine whether astronauts can assemble structures in space too massive for handling on Earth. In one test it was shown that a space-suited astronaut in zero gravity will be able to manipulate a component that would weigh 8,100 kilograms on Earth. Such tests are important in determining whether the huge orbiting structures that are now being proposed—from solar-power satellites to space colonies—can really be constructed in space.

EMERGENCY TRAINING

The technique of simulation touches other aspects of society as well. It has been suggested that if the power-plant operators involved in the two major blackouts experienced in the Northeast in recent years had had emergency training, the disastrous effects might have been averted. The problem, of course, is that utility-company personnel can't practice emergency procedures in a real-life emergency. The solution is a mockup of the utility control room in which the emergency can be simulated.

Now there are sophisticated control-room simulators for all kinds of utilities, including nuclear-power plants, in which operators can learn how to deal with even the most unlikely catastrophes. Simulators are also being used for air-traffic-control centers, for oil refineries, and even for the huge oil-platform drilling rigs used in the North Sea and elsewhere. Carriers for liquid gas cost about $180 million apiece. Handling gas cooled down to -162°C (-260°F) is both tricky and dangerous. A complete simulator has been built to train operators safely.

Another type of physical simulation is seen in the re-creation, on a reduced scale, of physical systems such as bays, rivers, dams, harbors, and so on. Models can be built for research as well as design purposes. "If we put a pier in the river here, what will happen there?" "If the rainfall increases by ten percent, how will that affect the water level in the bay?" Recent stringent environmental regulations require an estimate of "environmental impact" before construction is begun. It is in the simulators that wild speculations are tamed into reasoned projections.

As long as the characteristics of a system or piece of equipment can be put into a computer, its activities can be simulated: Changes in the ozone layer, the growth of weeds, world population forecasts —all can be simulated by computer. One interesting application is the projection of how structural frameworks will react to distorting forces, such as crashes, overloads, etc. The structures can include anything from thin experimental dams to automotive parts to the human skull.

* * *

SIMULATION SAVES

But with all this, the main application of simulators remains, so far, confined to aircraft. The reasons are many, but most important are the high cost of jet fuel, congested air lanes, and the fact that operating a modern jet aircraft can cost thousands of dollars per hour. It is obviously impractical to have a pilot practice takeoffs and landings in a big jet.

Al Ueltschi, president of Flight Safety International, estimates that his company, by training pilots on its 35 simulators—as opposed to "boring holes in the sky"—saved 95 million liters of fuel in 1979 alone. We may yet see the day when all training and certification of airline and military pilots are accomplished on simulators.

In general aviation there is: as in automobiles, an additional and very serious safety problem. Russell Munson reports in *Flying* magazine: "Pleasure flying, the worst category, is, by some calculations, ten times more dangerous than driving a car. That's unacceptable."

Indeed it is. Better flight training would surely help. Though costs of advanced simulators (they can run up to several million dollars) are still too high for widespread use in general aviation, wider use and falling computer costs may turn this around. Ueltschi estimates that the current breakeven point is $1 million. That is, unless an aircraft costs more than that, it's probably not worth making a simulator for it.

The main point is that simulator training is in fact better in many ways than actual flight time.

The instructor, for example, can freeze the flight at any point if he sees the student pilot doing something wrong. It thus becomes possible to "back up" and try again. If landings are a problem, it becomes possible to practice only landings, without takeoffs.

ILLUSION AND REALITY

During your next flight on a commercial airliner you might reflect on this: The pilot may have had no more than two hours of actual flight time on that particular kind of craft. If he's a Pan Am 747 pilot, he may *never* actually have flown such an airplane before. Pan Am has set up a test program, with the cooperation of the Federal Aviation Administration, in which, after 23 hours in a 747 simulator, a pilot's first flight is a regularly scheduled one —with passengers. But don't be too alarmed—during the first 25 hours of actual flight time the right-hand seat is occupied by a check pilot who is well experienced in the 747 and who keeps a watchful eye on the proceedings. Whether this requirement will be waived or decreased in the future remains to be seen.

Since the simulator itself is such an expensive machine, beginning pilots start out in simpler simulators—a cockpit-procedures trainer and then a flight-instrument trainer —before graduating to the real simulator.

A cockpit-procedures trainer, for instance, may be only a cardboard mockup of a real cabin, but it can familiarize the student with the placement of the instruments and controls.

Flight-instrument trainers usually do not have the visual capability of the big simulator and may have only two or three degrees of motion capability, instead of six. The old, world-famous Link trainer—which celebrated its fiftieth birthday last year—was just such a device. The lack of visual presentation isn't as much of a loss as you might think, for much of the most dangerous flying takes place when the pilot *can't* see anything, because of fog, clouds, rainstorms, darkness, and so on.

In one of the more spectacular examples of simulation, namely dogfighting in the air, two pilots work out in two separate 12-meter domes. Each sees the other's plane in his windscreen, and each is able to spin, dive, and do all the aerobatics necessary to bring himself into the best position for attacking the adversary aircraft, or to dodge its attack. The interaction is totally real. Every movement of the controls in one craft produces a corresponding change in the attitude and position of the craft as seen in the window of the opposing craft and as revealed on the instrument panel of the plane making the diversionary move.

Marc Liebman, an experienced ex-Navy helicopter pilot who has flown one of these craft in such a dogfight says he came out of the experience in a sweat. Yet at no point did either "aircraft" move so much as a centimeter.

If simulation becomes more prevalent in our lives, it will be due to work being done in the aircraft-simulation field. Here is the cutting edge of simulation technology. The software requirements alone for meticulous simulator images can be tricky. Highly accurate models of terrain may be built and a specially adapted television camera moved around above the landscape in synchrony with the vehicle being simulated. As you bank into a left turn, the camera tilts accordingly so that your view out the window is just as it would really be.

This model-board approach has largely been superseded by the newer computer-generated image (CGI) devices. Here, the miracle of the computer really shows. For once the parameters of, say, an airport and its surrounding terrain have been entered into the computer's memory, it can create an image of that terrain from any angle and change it appropriately as the craft moves. The original CGIs had only nighttime capabilities, the points of light being relatively simple to accomplish, but present models can produce full-color daytime images as well. The daytime images are still not fully realistic; they are stylized, resembling illustrations.

Fortunately, it turns out that airports, which have lots of geometric features, such as parallel lines, flat surfaces, lights, and so on, are simulated very well, while trees and other such features are harder to do. This works out satisfactorily because pilots of high-speed aircraft see little detail on the ground between airports anyway.

I should point out that even though the images are generated on flat picture tubes, this is not the impression one gets. By means of special optical devices, an amazingly realistic three-dimensional image, in full color, is created as we look out the windows. With the ever-increasing capability of computers, the images are becoming more realistic every day. It is likely that as the optical process called holography becomes more refined, the illusion of reality will be just about perfect.

In the simulated society of the future, make-believe will more closely approximate the real. Whether in learning how to defuse a bomb, avert a nuclear accident, breed a new strain of bacteria, play golf or tennis, design a new crash helmet or submarine, or perform a delicate lens-implant operation, we'll learn to save time, money, and lives by faking it.

BIOCHIP REVOLUTION

By Kathleen McAuliffe

While microchip architects race to squeeze more and more information onto wafer-thin silicon, a few pioneering biochemists are plotting a computer revolution that could make obsolete the most advanced circuits dreamed up in the back rooms at Intel and Motorola. Almost unnoticed, the ultimate biological computer has reached the drawing boards.

The bioprocessor will be a molecular latticework that can grow and reproduce. Capable of logic, reason, perhaps even feeling, its three-dimensional organic circuitry will not process data in the rigid, linear style of earlier computers, but network-fashion, like the living brain. Small enough to mesh directly with the human nervous system, biochip implants may restore sight to the blind and hearing to the deaf, replace damaged spinal nerves, and give the human brain memory and number-crunching power to rival today's mightiest computers.

The prototype is taking shape not in the high-tech labs of IBM or Bell, where researchers have focused on such hardware as optical disk memories and Josephson junctions, but at EMV Associates, Inc., an obscure six-employee outfit in Rockville, Maryland, where corporate gene splicers seem to be proliferating as quickly as their bacteria. EMV's president and co-founder has high hopes for his company: "Our aim is to build a computer that can design and assemble itself by using the same mechanism common to all living things. This mechanism is the coding of genetic information in the self-replicating DNA Double helix and the translation of this chemical code into the structure of protein."

A tall, burly Irishman with a hearty laugh and a Billy Carter smile, James McAlear manages to be both a shrewd businessman and a highly imaginative inventor with several patents to his name. He is both an idealist bent on expanding the human intellect and a pragmatist concerned with the nuts and bolts of how things work. Among his colleagues, Dr. McAlear is known for his eccentric and electrifying ideas and an irreverent wit born of childhood rebellion against a strict Catholic upbringing.

"It's easy to build a whole religion around this artificial-intelligence work," he says with a mischievous twinkle in his eyes. "After all, we are looking at conductive velocities about a million times faster than nerve cells, circuit switches one hundred million times faster than neuronal junctions, or synapses, and packing densities of the functional circuit elements a million times greater than are formed in the brain.

"This factor of ten to the twentieth power is truly incomprehensive in terms of any present concept of intelligence. It would be expected that the 'being' of an individual so equipped would live in the computer part, not in the central nervous system. It is also possible that when the corpus perishes, its implant would survive and could be transmitted to a fresh host. Well, that pretty much fits the specifications for an immortal soul. And if you have something that has intelligence and the ability to communicate at high speed, it might well become a single consciousness—a superior, an omnipotent being."

If all this sounds like the next sequel to *2001*, it is a fantasy that big business takes seriously. Bethesda Research Laboratories has just bought 30 percent of the company's shares. McAlear reports that DNA Science, Inc., the venture-capital arm of E.F. Hutton, and Paribas, a large French financial firm, are also keen to invest. The Japanese, too, are hot on the trail of molecular electronics. EMV is now arranging a deal with Mitsui Corporation, a Tokyo-based trading company that already has acquired a large stake in both biotechnology and microelectronics.

EMV has also been awarded a grant from the National Science Foundation to test the first living interface between an electrode-studded chip and the brain. If successful, the experiment could restore sight to the blind. Its success would also reassure McAlear and many other bioengineers that embryonic nerve cells can serve as a living bridge between the mind and the computer. It is here that the new technology may pay off first.

A little over a decade ago British scientists reported that a blind woman could see phosphenes, bright flashes of light, when they touched her exposed brain with an electrified wire. Since then, blind patients with electrodes implanted in their visual cortex have been able to recognize simple shapes, letters, and even short sentences in Braille. Impressive as it is, this vision is crude, like a cartoon sketched out in flickering light bulbs. Building on this work, McAlear hopes to produce an image with roughly the resolution of a staticky black-and-white TV.

The major stumbling block has been making the electrodes. Even hair-thin wires are too thick to link with only a single neuron; each stimulates a bundle of neurons to create a phosphene. McAlear intends to cover his electrodes with a layer of protein, drilling thousands of holes through the coating with an electron beam. The holes will be covered with polylysine, a cellular glue, and be attached to embryonic brain neurons. According to plan, the nerves on the implanted electrode will then grow into the visual cortex of the brain, forming links with individual neurons in the patient's cortex. Each electrode, McAlear projects, will produce up to 100,000 phosphenes, 6 million in all.

"The idea is to mount a tiny video camera on the

glasses of a blind person," he explains. "The wires leading from the camera will be inserted through a tiny plug in the scalp and connected to the protein chip and its electrodes. The computer will then process visual data and convert them into a pattern of phosphenes that will duplicate the camera's image."

If all goes well, EMV expects to have a prototype ready for testing in ten years.

At this point, however, the company is clearly a gamble for its eager investors. McAlear's organic circuits will measure less than one hundredth of a micron across. (For comparison, a human hair is about 62 microns wide, a red blood cell, 7 microns.) Building them will not be easy.

When approaching the limits of smallness, solid-state physicists enter a world as alien and uncharted as the universe. Just as black holes warp space and time, the infinitesimal distorts reality as we know it. Here traditional notions about the nature of solids are turned upside down. Electrons pool at discrete energy levels, turning conductors into insulators. Here a stray cosmic ray can knock an electron off course, causing malfunctions in software or, worse, damaging the molecular hardware. Computer scientists, genetic engineers, organic chemists, and a host of other specialists will be needed to make the biocomputer work. McAlear has managed to attract these diverse talents—probably, he thinks, because of his eclecticism.

This "eclecticism" is reflected in his life-style as well as his professional creativity. His office resembles a baby's playpen. The coffee-stained papers on his desk might have been blown there by a tornado. A file drawer opens to reveal balls of crumpled paper. Underneath this apparent chaos lies not a hint of order.

Yet this inability to compartmentalize may well explain his knack for seeing relationships in seemingly unrelated findings. In the junk shop of his mind, ideas mingle freely, coming together in unexpected combinations. "The inspiration for a molecular computer," McAlear comments, "would have occurred to anyone who had bothered to think about the ways in which biological principles might be applied to the manufacture of integrated circuits."

It started with the notion of building a conventional chip out of protein. Proteins organize the vast array of biochemicals and assemble them into living organisms. And if they could create order from that incredible complexity, McAlear reasoned, surely they could serve as a simple matrix for microcircuits.

To test this idea, EMV's pilot study coated a glass slide with a monolayer of protein, which was in turned covered with a thin protective material, called a resist. Then an electron beam was used to dig trenches in the resist about half the width of a red blood cell. Dipped into a silver solution, the protein exposed by the beam organized the metal into fine streaks—microscopic wires. The research, carried out with Professor Jacob Hanker, of the University of North Carolina, in Chapel Hill, produced the basic pattern of a functioning microchip.

Dr. Hanker believes that proteins can arrange nearly any metal into useful circuits, including many materials that cannot be used by conventional methods. And protein can be coated with protective layers thinner than silicon. This makes it possible to etch much finer lines than those used in today's chips. Small-enough conductors could pack onto biochips more than 100,000 times the power of present computers. Gentronix, EMV's holding company, intends to test such a "very small device" (VSD) within three years.

The VSD could soon stand the computer industry on its ear. But in McAlear's grand scheme it is merely the first step in the development of an ideal biocomputer—one modeled entirely after living systems.

Living organisms are treasure houses of molecular conductors and switching circuits. Poorly understood, their components are far smaller and more densely packed than the electronic components of the most sophisticated computers. Take, for example, the electron transport chain responsible for photosynthesis. The leaf of a green plant contains 10 million more electronic elements per square millimeter than a silicon chip. Without high-powered electron beams, nature has succeeded in building a microdevice that rivals man's most prized intentions. What is her secret?

As one scientist explains, "Nature uses a bottom-up, bricks-and-mortar method of building up microstructures atomic layer by atomic layer." This is how McAlear plans to build his biocomputer. Like many natural proteins, the molecular computer will be grown from the DNA templates of genetically engineered bacteria. The result will be radically different from the VSD, where protein merely supports traditional metal circuits. McAlear's brainchild is a three-dimensional protein lattice, including its densely packed circuitry.

"This microscopic gem will contain the collective consciousness of mankind," he envisions, ticking off the advantages of molecular computers. "They open up the possibility of three-dimensional circuits, increased speeds, reduced energy consumption, and ultraminiaturization that can reach a million billion elements per cubic centimeter. On this scale, all the memory elements of every computer manufactured to this day could be contained in a cube one centimeter on a side."

Since the biochemical organization of living cells is very different from what would be desirable in a molecular electronic device, an obvious question arises: Where in nature will genetic engineers find the genes that code for these proteins?

Very simply, they won't. Biotechnologists need not make do with the hand-me-downs of evolution. Scientists have already begun to design synthetic proteins unlike any found in living cells. Computer graphics allow them to study the three-dimensional structure of molecules that exist only in theory. Soon these computer techniques will be coupled with recombinant DNA technology to make the genetic blueprints for make-believe proteins.

"The ultimate scenario," says geneticist Kevin Ulmer, of Genex Corporation, "is to develop a complete genetic code for the computer that would function as a virus does, but instead of producing more virus, it would assemble a fully operational computer inside a cell." Eventually, he thinks, this will be a large market for the gene-tampering industry.

Biobuilding microcircuits would be largely a self-correcting process, Ulmer notes: "One component could not assemble out of place or out of turn, because it would lack the necessary binding sites required of the correct molecule. The yield of perfect devices could approach one hundred percent." And every time the cell divided, a new assembly line would be born. Conventional factories will have a hard time competing with microbes.

It will be a long time before working computers roll out of their cellular factories. "Ulmer's scenario is at least thirty years off," predicts Dr. Zsolt Harsanyi, vice-president of DNA Science. "Computer graphics are very useful, but we will need a far greater knowledge than we have now to design proteins that will automatically self-assemble the way a virus does."

Given these challenges, Dr. Forrest Carter, of the Naval Research Laboratory, in Maryland, has begun to look for solutions today instead of, as he puts it, "waiting for the rope to run out in the semiconduction field." A tall, bearded man who would look at home in the British Admiralty, Dr. Carter is designing molecular switches to record the on-off binary code used in computers. "One possibility," he says, "is to use long chain molecules with alternating double and single bonds. When an electron is transported down the chain, all the single bonds will switch to double bonds, and vice versa. This, in turn, will determine whether intersecting chemical chains are switched on or off."

A computer built with Carter's molecules would be so small that an electron traveling from switch to switch would have almost no chance of colliding with an atom or another electron. So it would produce almost none of the waste heat that sometimes causes today's computers to fail. Its speed of operation would closely approach that of the superconducting Josephson junction, on which IBM has pinned its hopes for a supercomputer, spending an estimated $100 million. And its power needs would be so low that the computer could run on the chemical energy of the cells around it.

The gemlike biocomputer of McAlear's dreams, implanted in the brain, will sprout nerve projections from its tiny protein facets. The host's neurons will link up with these spindly outgrowths, sending out electrochemical pulses in the brain's own language. Inside the molecular lattice, electrons will dance at the speed of light. Only at its periphery, where the implant's tentacles sense the neighboring nerve cells, will the flow of electronic information slow to the gentle pace of the brain.

Suddenly the old distinctions begin to fade; the rigid line between life and nonlife wavers. How do we classify a human invention that derives its materials and energy from the living cells around it and can grow, reproduce, and think?

"By definition," says McAlear, a Harvard-trained biologist, "this is a form of life, but its potential will be realized through research and development by human beings rather than the trial and error of natural selection." The biocomputer is not just another implant, he feels, but a symbiote, living from the cells it inhabits and giving them, in turn, the chance to evolve into a higher intelligence.

Evolution set the precedent for this symbio-genesis some 3 billion years ago, McAlear says, when microorganisms assimilated even more primitive creatures that now survive as the chloroplasts, mitochondria, and other organelles of modern cells. "This alliance," he says, "paved the way for all plant and animal life. But the next time around, symbiogenesis may lead to intelligent beings that would be superlife by present standards."

The implant McAlear envisions would ideally combine the brain's ability to relate incoming data—to reason—with electronic speed and efficiency. "Computer and human intelligence are not antagonistic," McAlear asserts. "They complement each other beautifully. You might consider it a marriage made in heaven."

McAlear and John Wehrung, EMV's co-founder, admit that so far they have no idea how to achieve such a marriage here on Earth. "First and foremost," says Wehrung, a computer specialist, "we have to figure out how the brain works and then strive to duplicate its mode of operation."

In sharp contrast to McAlear, Wehrung is neat and athletic, fair-haired, with boyish good looks. But, like his partner, he loves to contemplate the bizarre. "The next plan of action," he says in hushed tones that sound almost intentionally theatrical, "is to develop a compatible symbiote that will literally grow into the brain, establish communication with individual neurons, and thus learn from them in a biological sense."

Farfetched as it sounds, Wehrung's idea might require only slight changes in the electrodes designed to simulate sight in the blind. Instead of sending signals into the brain, the device would eavesdrop on the electrical dialogue between cells. "It would infiltrate the tissue," Wehrung says, "live with the tissue, sense the tissue, and keep detailed records of its interaction. After all, how do you learn how the natives live in Africa? You can take pictures of them. You can climb a tree with binoculars and gaze at their activities from afar. You can study the artifacts of their culture. But the best method is to send an anthropologist into the jungle to befriend them, live in their huts for a year, take part in their rituals, and report back. A symbiote would serve an analogous function."

McAlear, for his part, is not one to downplay the importance of the biocomputer he and Wehrung will build once they have the symbiote's report. "We are so accustomed to thinking of ourselves as the crowning glory of evolution," he says, "that it is difficult even to consider the possibility that we are merely the beginning of life—a potential for intellectual development that is limitless once we take control of our biological density.''

His voice rises slightly to penetrate the din of the bar in which he is speaking. "One Sunday about four years ago," he begins, "I was parked in front of the Swedish Covenant Church, waiting for my wife, Anna, who was inside. The minister, it seems, had gotten into this tiff of singing a hymn and then praying that I'd come in and be saved. Finally my wife came out. 'Why don't you come in,' she implored. 'They've asked me to ask you, don't you believe in God?'

" 'You go back,' I said, 'and tell them that not only do I believe in an almighty God, but I'm probably the only one here that has any idea how to build it.' "

PART THREE:
SOFTWARE

INTELLIGENT MACHINES

By Thomas Hoover

HAL, the sentient computer of *2001: A Space Odyssey,* murdered most of the spaceship crew it was built to aid and protect. The idea of a superintelligent machine turning on its human masters is one of the oldest in science fiction. But science fiction is finding it increasingly hard to stay ahead of science fact, and nowhere is this more true than in the field of machine intelligence. For example, how far into the future would you place beneficient, seemingly intelligent computers that:

- Converse with people over the telephone, using properly grammatical synthesized speech;
- Interview a physician about a patient's symptoms and either make a diagnosis or recommend further medical tests;
- Clean up in a poker game by analyzing each human player's style to decide when to bluff and when to call or close out their hand?

If you plunked these scenes squarely in the present, you're right. Our smart machines are still a long way from seeing, speaking, and scheming, as HAL did, but they do exist, and they're beginning to change our lives. They're all part of the world of artificial intelligence (AI), described by MIT's Marvin Minsky, one of its founders, as "the science of making machines do things that would require intelligence if they were done by men."

We have always assumed that humans alone are capable of unlimited speech. We have credited this unique ability partly to our physical design and partly to a brain that can handle abstractions. Running a slow third comes the sometimes questionable assumption that we have something to say. It appears, however, that remembering and reproducing words are not really all that hard for machines, either.

Computers, it turns out, can talk very nicely. Researchers have figured out how to translate sounds into digital form—a technique that has many uses. Since strings of numbers can easily be stored on an integrated-circuit chip, computers can now pack away an extensive vocabulary. Talking timepieces and hand calculators are already on the market, and Texas Instruments has brought out Speak and Spell, a device that pronounces words and asks the user to type in the spelling. Speak and Spell stores over three minutes of speech on two tiny memory chips, and two new language translators can spout some 500 words. The inexpensive storage and processing chips developed for these devices underscore the imminence of widespread talking computers.

Repeating words doesn't take all that much intelligence, but what about reading aloud? If a computer could pronounce words that weren't already programmed into it, would that show intelligence?

Consider the Kurzweil Reading Machine. It can scan a page, recognize letters and words, apply phonetic rules, discern phrase boundaries from syntactical and vocabulary knowledge, figure out where stress and accent should go, and finally synthesize it all into words and sentences as it goes along. Priced under $20,000, the device has been a great help to the blind. What's more, if given a little practice, people find it just as understandable as a human reader's voice.

The Kurzweil machine uses the same knowledge of pronunciation we do (whether we are aware of it or not) to turn print into speech, and it performs about as well. Synthesized speech is one of the more successful areas of AI research. And yet, impressive as the Kurzweil device is, its performance still falls far short of true intelligence.

Like many people, computers find it harder to listen than to talk. So far, most commercially available devices force you to pause slightly between words so the computer can detect word boundaries. Almost all computers must get to know you. You have to train them by repeating each word several times to form a personalized "template" of your voice characteristics. Thereafter the computer simply compares each word with the vocabulary in its memory until it finds an acceptable match.

Word-recognition devices are faster and more congenial than standard computer terminals, and new uses for them are springing up all around. People are now telling computers to sort packages, fill out forms, and keep track of stock-and-bond transactions. A system now being tested will enable pilots to talk with in-flight computers. But perhaps the most rewarding application will be their use as an aid to the handicapped, allowing people without the use of their hands, for example, to tell a robot arm to do their bidding.

The next step beyond "discrete-utterance recognition"—and it's more like a giant leap—is the computer's ability to understand continuous speech. When you want a computer to pick out individual words in rapid-fire colloquial speech (we don't really say, "Did you?" We say something like "Didja?"), the machine has to start looking beyond mere acoustic matching. Enter artificial intelligence.

One approach is to teach the computer to figure out what you probably would have said by using its understanding of grammar, context, and logic. AI researchers call these three types of computer knowledge syntactic, semantic, and pragmatic.

Scientists at Bell Labs are trying to produce a computer that can take airline reservations over the phone. Their work provides a good example of how this secondary information can be used. When you ring up the airline computer, a synthesized voice asks for your identity number and poses a series of questions, including where and when you wish to fly.

Pragmatic knowledge tells the computer you will be talking mostly about places and dates. So it doesn't have to compare questionable nouns against everything in the dictionary. Semantics tells it that when you say, "The day I want to travel is mumble," the mumble isn't going to be the name of a city. Syntax dictates that after the verb you will probably use a noun, a preposition, or an article, not another verb. In noisy surroundings we often rely on such cues ourselves.

Using these techniques, the computer can apply its sense of what you probably said to help sort out and identify the individual words. Since our mind most likely does something similar, this simplified linguistic analysis represents a kind of low-level artificial intelligence—probably the only true "intelligence" achieved to date.

Where is this route to AI leading? Well, work at IBM's Yorktown Heights Research Center, in New York, may produce a "voice-driven" typewriter. The research at this lab is by far the most ambitious in the field. IBM scientists are trying to recognize natural grammar-based speech, using concepts from information and communication theory as well as acoustic processing. IBM's speech-processing consultant, Rex Dixon, says that IBM's is the only research now grappling with totally unconstrained, natural language. All other work has been on grammars contrived to aid speech recognition.

Pressed to predict when voice-driven typewriters will appear, Dixon ventures, "If funding is maintained at a fairly high level, we can reasonably expect to have usable continuous-speech-recognition systems working with all talkers, natural grammar, and large vocabularies in fifteen or twenty years." In case you're counting, that's before the year 2001.

As desirable as spoken input is, all it really amounts to is a new medium for the same message. As long as we talk with computers on their own terms, we'll always be limited in our ability to communicate with them. So why don't we just teach them to understand everyday English?

That innocent-sounding question catapults us out of applications and back into academia's basic research, where AI experts have been stymied by the complexities of language. The problem is that natural languages tend to be ambiguous, something we instinctively compensate for with our firsthand knowledge of the world.

When we say, "I drove the car into the garage, and it died," we know that "it" refers to the car, not to the garage. To interpret this sentence as easily, a computer would have to know something about cars and garages. Understanding requires at least some knowledge of the subject.

Because AI researchers couldn't wait for the development of a worldly-wise computer, they have created some clever ways to get around the problem. Joe Weizenbaum, of MIT, hit on the idea of teaching a computer to bluff its way through a conversation by assuming the role of a "nondirective" psychotherapist, leading patients to understand their problems by asking for repeated explanations and clarifications.

In the exchange below, the computer is interviewing a troubled young woman, who began with a common complaint.

Human: "Men are all alike."
Computer: "In what way?"
Human: "They're always bugging us about something or other."
Computer: "Can you think of a specific example?"
Human: "Well, my boyfriend made me come here."
Computer: "Your boyfriend made you come here?"
Human: "He says that I'm depressed much of the time."
Computer: "I am sorry to hear that you are depressed."
Human: "It's true. I am unhappy."
Computer: "Do you think coming here will help you not be unhappy?"
Human: "I need some help. That much seems certain."
Computer: "What would it mean if you got some help?"
Human: "Perhaps I could learn to get along with my mother."
Computer: "Tell me more about your family."

The program has mastered the vocabulary of psychiatry just well enough to pick up on key words. For example, it recognized "depressed" as an undesirable state and knew that "mother" referred to family. When it draws a blank, it simply changes pronouns and rephrases the statement as a question, with an implied "hmm."

Weizenbaum intended to use the program for language research, and he was disturbed when people started taking it seriously. Some even demanded to be alone with the terminal in order to pour out their troubles. His dismay at human gullibility transformed him into a crusader against the potential misuse of AI.

Terry Winograd, now of Stanford University, takes a completely different approach to the problem. Early in this decade he devised a program capable of extended conversations by the simple expedient of giving the computer something to talk about. He created a "microworld" peopled by colored blocks, pyramids, and boxes on the computer's TV screen. He told the program everything it needed to know about this world: the relative size of the objects, their shapes, colors, and so on. He could then command the program in simple English to move the items around on the TV screen with a simulated robot hand. (He also taught the program enough grammar for it to reply in reasonably high-quality typed English.) A sample:

Human: "Pick up a big red block."
Computer: "Okay."
Human: "Find a block which is taller than the one you are holding and put it into the box."

Computer: "By 'it,' I assume you mean the block which is taller than the one I am holding."
Human: "Okay."
Computer (hand sets down the block it is holding and picks up a taller one, putting it into the box).
Human: "What does the box contain?"
Computer: "The blue pyramid and the blue block."

It's hardly something we'd care to discuss over brandy on a cold winter evening, but at least the computer had full command of its topic.

Winograd's work proved that computers could carry on a conversation on a limited subject, and language research blossomed in several directions. UCLA psychologist Kenneth Colby followed Weizenbaum's techniques, but he decided to take computer therapy seriously. To see how a computer might interrogate a patient with mental problems, Colby turned the experiment around. In the early 1970s he created a program that could converse in English and respond to questions as if it were a human paranoid.

Called PARRY, it was given the identity of a twenty-eight-year-old man who lived with his parents and bet heavily on horses. PARRY was obsessed with the idea that he was being stalked by a vengeful bookie. In one experiment six psychiatrists interviewed both PARRY and a real paranoid through a terminal. Asked which was human, they guessed wrong often as not.

Interesting as these achievements are, they are all

microworld research that deals with restricted domains of knowledge and limited vocabularies. The question is how to extend this to the real world.

Roger Shank, of Yale, is teaching computers "scripts" about human social situations, hoping to learn how we draw on expectations to extract implicit meaning from ambiguous statements. One of his scripts enables a computer to answer questions about what happens when it goes to a restaurant, recalling that paying the check presumes having eaten, and so on.

Other programs developed at Yale can scan a newspaper article, summarize the major points, and then update the summary daily as new details unfold in the press. Whether or not this research will someday lead to the use of stored knowledge for understanding language, it's now only a simplistic model of human cognition.

Ten years ago AI researchers thought that if they could teach a computer to talk about some artificial situation, they could then extend this language ability to the real world. Now they are less sure. The issue, once again, is context and experience. One often-repeated story tells of a machine that translated the biblical verse "The spirit is willing, but the flesh is weak" from English to Russian and back again. The computer came back with "The wine is agreeable, but the meat has spoiled."

We also acquire much of our knowledge visually, and if computers are going to equal our intelligence, they must do the same. The punch line here is that visual perception may be even more complex than language. It's easy enough to dump visual information into a computer. Any color-TV camera can easily pick up a deluge of data. The problem is in what happens next. No one understands how the brain makes sense of the information our eyes supply. How can we effortlessly process a jumble of light, shadow, movement, and color to recognize that the car speeding past is this year's new Jaguar, even when we have never seen one before?

The information continuously processed by the eye is so far beyond any computer's capacity that AI researchers have begun to suspect that the eye and brain somehow cheat. Perhaps they reduce the data to manageable proportions by pulling some trick to avoid processing redundant stimuli.

In hopes of simplifying the visual information a computer must deal with to digest an image, MIT's David Marr has tried to convert an incoming picture into what he calls a primal sketch. The light intensities in a real image are converted into a "rich symbolic description" of the way they change over a visual field. By remembering only where the change occurs, rather than everything in its field of view, the computer can cut its data processing markedly.

Minsky thinks our brain may store a number of handy reference pictures, which he calls frames. Suppose, he says, we are standing in a living room we have never seen before. We already have a rough idea of what we'll find in a living room and where things will be. So we don't have to start from scratch and ponder everything in detail.

Not surprisingly, research in computer vision inevitably leads us back to the matter of human experience. As in language programming, academic researchers have found that human intelligence is intimately tied with human perception. It's going to be very hard to duplicate one without also duplicating the other.

We may never need to mimic the eye's exact mechanisms, however. Our eye is a general-purpose instrument that may work as it does only because it had to grow from a single cell. Computer vision systems may someday surpass the eye's data-gathering ability precisely because they do work differently. Industrial vision systems already improve on human performance in that they don't get bored or show up at work bleary-eyed from a hangover.

In the early 1960s scientists at Johns Hopkins University created one of the first mobile robots, affectionately known as the Hopkins Beast. Said to resemble a garbage can on wheels, it was entirely self-contained, with no external computer link. Its sole activity was to whir up and down the corridors of its building in a sonar search for electrical outlets. Whenever it spotted one, it would home in, screech to a halt, insert its plug, and feed. When sated, it pulled away and began its quest for the next outlet. It had eyes for only one thing: electric-outlet coverplates. Its search for "food" was the one "instinct" that kept it going.

SRI International, then known as the Stanford Research Institute, built a more sophisticated mobile robot in 1968. Dubbed Shakey because it always seemed unsure of its footing, it was about the size of a small desk-top copy machine, perched on a mobile base. Its "head" was a TV camera and range finder, topped by a radio antenna. Guided by radio from a large nearby computer, Shakey could decipher English commands and figure out ways to obey them. Told to retrieve a box sitting out of reach on a platform, Shakey was smart enough to survey the room, find a ramp and shove it against the platform, then motor up to claim his objective. He could see, move, touch, and reason.

Unfortunately, Shakey never learned to leave his microworld playpen. His intelligence was far too limited for him to roam abroad. The conclusion that unlimited mobility requires human sophistication halted university research on sentient robots.

Robot research now rests mainly in the pragmatic hands of industry, which asks only that machines be smart enough to handle their jobs. Although manipulators and so-called blind robots are nothing new to the industrial scene, until recently they had no sight, no sense of touch, and no "brain." Without these they will cheerfully screw a bolt into thin air if somebody forgets to put the socket in exactly the right place.

This is all about to change. General Motors' vision-equipped manipulators, called hand-eye machines, are now entering the assembly lines. The GM system identifies unoriented parts as they come down a conveyor belt and maneuvers a hand to pick them up and fit them to a car body or discard them into a reject bin if they're not the right size. A robot system recently developed at

SRI International sports both eyes and a sense of touch, enabling it to sense when a bolt is not screwed in properly. Vision-equipped robots are slowly going to revolutionize the assembly line. Kawasaki Ltd. is even planning factories run entirely by robots.

Charles Rosen, SRI's former director for robotics, says, "Although ten to fifteen percent of assembly jobs will still be done by people for many years, a heck of a lot of other jobs can be done by robots." He thinks robots will eventually cut the workweek, shift people into service jobs, and make low-cost custom products possible.

For years AI's most widely known use has been in computerized chess. Though machines must struggle to compete with human sight and language, they have much less trouble with analytic manipulations. Grand Master David Levy recently won his ten-year-old bet, made against several AI experts, that no computer chess program could beat him by 1978. (See Interview, April 1979.) However, Levy took only three out of five games against the program "CHESS 4.7," and he is not planning to renew his bet for another decade.

Computers became unbeatable in checkers several years ago, and Nicholas Findler, at the State University of New York at Buffalo, seems well on his way to developing a program that can beat most poker players, regardless of their style. Poker is particularly interesting, since winning demands a lot more than simple logic. Human psychology is the key to this game. Significantly, the program's opponents, isolated and playing through a keyboard and TV screen, often cannot tell which of the other players is the computer.

Games may seem trivial, but scientists believe they provide important models of human problem solving—a major goal of all AI research. But if computers can play championship chess, mustn't they surely be capable of simulating the human problem-solving process? Unfortunately, no.

Quite simply, no one knows how we acquire our "common knowledge," or how we relate it to new situations to produce "understanding." Some researchers are now concluding that intelligence is so entwined with life itself that it can never be fully reproduced in a box of silicon chips.

The academic debate over what constitutes human intelligence is far from settled. AI researchers interested in practical problem solving spend little time worrying about it. And although these uses haven't taken the world by storm, they have shown that computers can do more than play chess.

The PROSPECTOR system at SRI aids in mineral exploration. A program at Stanford, called DENDRAL, is being used to deduce molecular structures from the output of mass spectrometers. At MIT, MACSYMA does high-level algebra for mathematicians. J. D. Meyers, at the University of Pittsburgh Medical School, is developing a diagnostic program called INTERNIST, which incorporates his own experience as a physician into a simulation of clinical "judgment." And at Stanford, MYCIN will soon aid in diagnosing blood infections and meningitis.

The thread that connects these programs is that they are restricted to a narrow range of facts. They are what Stanford's own Ed Feigenbaum calls "knowledge engineering." They do not pretend to be exact replicas of human thought processes. Rather, they use the computer's vast memory and high speed to sift through what for human beings would be an overwhelming mountain of information. We avoid being overloaded with data by using judgment to focus on what is important. Thanks to their speed, computers can arrive at the same result by looking at all their data every time. They may appear "intelligent," but they tell us very little about how the human mind works.

So far, AI research hasn't even come close to reproducing human intelligence. AI has painfully discovered that our type of intelligence cannot really be isolated from the rest of human life and experience. Attempts to process visual images, create language, and solve problems have given us a new appreciation of just how complex the brain's functions must really be.

Yet machine intelligence, with computers doing things their own way, has almost limitless potential. Sooner or later computers will probably duplicate most everyday human tasks. An artificial intelligence is indeed arising, but it is a species different from human intelligence. It is evolving rapidly and, despite its alien character, is adapting nicely to man's world. We have yet to face the psychological or social implications of its Faustian success.

ILLUSORY SOFTWARE

By Robin Webster and Leslie Miner

Since advances in robotics and bionics have led to a deeper understanding of how our bodies work, psychologists and computer scientists have long hoped that attempts to create an intelligent machine would help us learn more about the human mind.

Unfortunately, such attempts are blocked by a double barrier. The first is sheer ignorance; we know so little about how our minds work that we can't possibly duplicate the process.

"Right now," says Dr. Richard Gregory, of Bristol University's brain and perception unit, "the whole thing is opaque. Suppose you are doing a job and you arrive at a decision. I don't know how on earth you did it. We don't even know how air-traffic controllers, for example, make decisions. You find in fact that there are big differences among individual controllers."

It is the hope of researchers like Dr. Gregory that by developing primitive thinking machines they may create an analog to the human process that helps to explain it. "It seems to me," Gregory says, "that the whole technique of trying to make the decision-making process more explicit in a machine, which you can, after all, monitor, is the way to approach the problem."

Here, though, researchers meet the second barrier: the fascinating probability that thinking machines, like humans, will be subject to illusion. The more original and creative intelligent machines become, the less reliable their final decisions will be. In fact, the more they seem to reflect our thought processes, the more they mirror the caprice of our judgments.

"I would define illusion as a discrepancy between a description and reality," Gregory explains. "In the case of perception, it is a discrepancy between what you see and what you believe to be the case conceptually.

"The hunch, therefore, is that if a machine is going to show originality and come up with novel solutions, then it is almost certain to be unreliable. This is because it's got to have the facility for getting outside its normal loops of operation. I don't think it can ever have an adequate set of rules to look for a novel solution. So if you ask a computer to do this, I think you are buying an increased probability of error."

According to Gregory, in the future we may have a virtual think tank of artificially intelligent machines supplying us with conceptual information that would otherwise have been beyond our reach. "We will use machines to seek out unknown realms, artificial or otherwise."

To deal with the probability of error, he sees the development of a new science of error. The whole nature of learning and knowledge would change in such a future. Some machines will deal with mundane matters, while others will exist in the fragile world of insight and illusion. In this way, Gregory believes, like the speciation of animals, there will be a speciation of thinking machines.

Machines used in wartime to aid decision making, for example, must have a totally different reaction to information than a novel-writing machine. In the creative machine, illusions would be acceptable, even welcome, but not in the military unit. "Machines used to aid decision making in wartime," Gregory says, "face a totally different level of illusion. Such systems have to detect aircraft and troops in the battle zone at the limits of their capabilities since they must be out of enemy attack range. The need for reliability in these machines is high, yet their decision-making processes are typically based on very inadequate information."

In this circumstance any degree of illusion could be disastrous. So military and creative thinking machines may ultimately become as distinct as generals and poets.

The illusion barrier will slow the evolution of intelligent machines for some time, Gregory feels. "Consider the law, if you like. You surely can't have rules for convicting or trying criminals as complex as the situations that people find themselves in, because the rules can't be as diverse as reality. Rules can reflect only a part of reality. As with the law, you are absolutely bound to get mismatches with machines."

If we do manage to give machines intelligence, will they become more human or will they develop a "machine psychology" different from our own.

"They will become less like computers as we understand them," Gregory suggests. "Personally I take the view that any machine is extremely interesting conceptually. I like the example of a sewing machine.

"Although it's just going up and down, it is actually carrying out an extremely clever bit of functional operation. In order to do this, very early sewing machines were just like fingers with needles; there is a series of them in the Smithsonian. As time goes on, however, things get less and less like the human being but still carry out the deep logical functions by a different mechanical means. I think this is a very good way to look at artificial intelligence. It doesn't have to look like a human being, but deep down the function should be similar."

Ultimately the process of machine decision making may to some extent, remain unfathomable. Gregory thinks, though, that we might cope with machine illusions by programming them to recognize the symptoms. He breaks decision-hampering illusions into three categories, incorrect assumptions, inappropriate strategies, and distorted input signals. "There are criteria by which illusions can be detected. It may be a difference between what a machine reports and what we see or what two machines report. Or it may be that a machine gives a dramatically different answer with a slight change in the problem.''

Gregory uses the example of a radio telescope. If its detection equipment malfunctions, signals will be distorted. If the procedures for gathering information through it are faulty, the input will again be worthless. Finally, if the telescope is pointed at the wrong star, the information won't be useful. Any of these could cause unreliable results in a thinking machine.

This raises the root question of whether a thinking machine is "conscious" or merely "intelligent." Gregory thinks we should carefully separate the two. In his view, intelligence lies somewhere between a Xerox machine (which is repetitious but lacking in novelty) and a random-number generator (which is all novelty without any purpose). However, he says, if someone stands on your pet dog's tail, you believe that it hurts the dog not because it is intelligent but because it is conscious.

"The question is philosophical," he notes. "Why are we conscious? Is consciousness necessary for high levels of intelligence, and, if so, is consciousness causal? If it is causal, how can we expect to find it in a computer program or an electronic system? It appears to be something categorically different from any formal process, and this is where we get into metaphysics. It is a deep problem to which we have no answers."

MIND MACHINES

By Philip J. Hilts

Ken Thompson was nervous when I arrived at Bell Labs for an interview. His computer had been arrested.

Belle is the world computer chess champion. Thompson had boxed Belle and was shipping it to the Soviet Union for an exhibition when Customs agents seized it on the grounds that computers going to the USSR might give away American technological secrets, especially secrets that might be used militarily. Thompson says the only way Belle could be used militarily would be "to drop it out of an airplane. You might kill somebody that way."

Thompson would get the machine back some weeks after my visit so that it could once again do battle in its own sphere of combat: taking on top chess players and, for the most part, vanquishing them. Not that Belle is mindful of its wins. Like all programs in artificial intelligence, it has no mind at all beyond its specialty. It doesn't know that it exists, or the name of the game it plays. It doesn't know its rights when arrested. Yet in the tiny fraction of the world where chess moves are important, it has extraordinary power. Programmers call the power "brute force."

The number of chess players who can beat the best computers is down to perhaps 300. The 30,000 other serious human chess players in the country cannot win against the modern machines. Thompson told an illustrative tale: At a chess festival in Hamburg a year and a half ago, he says, the chess grand master Helmut Pfleger gave an exhibition of simultaneous play—walking from board to board and playing 26 opponents at once. When it was over, he had won 22, drawn 2, and lost 2. He was asked afterward whether he had suspected that there was anything unusual about those who played against him.

No, he had not. Then he was told that three of his opponents had radio receivers in their ears and the moves they played were dictated to them from three computer chess programs. One of the three was Belle. When he learned that one of the players who beat him actually was Belle disguised as a human, Pfleger was "upset, really shocked," Thompson says.

Grand master Pfleger was not the only one who couldn't distinguish between human and computer play. After the match, five games from the exhibition were distributed to chess players around the world. They were asked to pick out the computer's game from among the four human and one computer games presented. In looking through the games, chess players, including grand masters, looked for blunders they thought would be made by a computer. On the whole, the chess players guessed wrong.

This little experiment may have had more significance than first appears. Early in the history of machine intelligence, mathematician Alan M. Turing recognized that there could be an endless debate about whether machines might be said to "think" the way people do. He devised a hypothetical game to settle the question, involving a person and a computer, each closed in its own room. An observer outside the two rooms, communicating only by teletype, would have to decide through dialogue which was the person and which was the computer. In the realm of chess, at least, a machine passed Turing's test.

It is interesting, though, that a few of the computer specialists who took part in the test picked out Belle's game. Their method was the opposite of the chess masters'. They looked for dumb moves that only a *human* would make. They were right; the best of the games was Belle's victory.

Do computer chess programs now face a barrier to future success, the wall that separates most players from the masters? "People always say things like that," Thompson says. "They have said it at every minute in the history of computer chess. But computer chess goes through those limits in absolute linear motion. I mean it just walks right through."

What Thompson is speaking of is the power of a machine brain, not a machine mind. The brain is a machine for calculation. Fed carefully pared and tenderized information, it is a paragon of brute logical, mathematical force. The mind can be viewed as a similar machine, but one susceptible to, and immersed in, the world, filled with billions of bits of knowledge and experience—of the magical, the irrational, the sensual, the honorable, and so on.

The great chess machines of today are wired to do nothing but extremely fast number crunching on problems that arise in the game of chess. These machines know nothing of wooden boards or ivory pieces or the right psychological moment to strike. They are not storehouses of chess knowledge and lore, as humans are.

The fact is, these powerful chess machines are fundamentally chess-stupid. They can't think up strategy or follow a plan of action from move to move. Some of the simple rules of thumb learned by novices are not known to the computers. Their "evaluator" functions, which decide the value of the moves they look at, are crude. So the burst of power that elevated the rank of computer chess machines has come as just as much a surprise to their programmers as to the human victims they defeat.

From the time when computer intelligence pioneers such as Claude Shannon and Alan Turing first considered the problem of how to make chess-playing computers, designers believed that sophisticated, humanlike knowledge of the game would have to be put into the machines to make them even passable opponents. The other alternative—merely letting the machine "look at" every move, all possible responses, all the responses to those responses, and so on—was thought to be a useless undertaking. After all, examining all the moves and their possible answers would mean looking at something like 38^{84} possibilities, most of them completely worthless. A computer operating at the speed of light would be billions of years at the task. Yet, so far, raw calculating power has carried the day. Machines don't examine every possibility, but they do explore millions of moves on each turn.

There was one moment, says Northwestern University's David Slate, cobuilder with Larry Atkin and William Blanchard of the computer chess program with the longest history of success, that indicated to him the emergence of computers as serious players. It was six years ago, at the Paul Masson American Chess Championship, in California. The tournament draws hundreds of players, from the bottom rank to grand masters, who enjoy the ambience: Boards are set up outdoors, and the sponsor serves wine between rounds.

At the time of the tournament, Slate's program was the most powerful then built. No one, including Slate, took the program very seriously as a contender in a human tournament. But in that tournament, for the first time, the program would be running on a very big, very fast computer—the Cyber 176 of Control Data Corporation.

Slate entered the B-class tournament. He thought the category might be too high for his program. He thought there was a good possibility that it would lose every game. And as the tournament got under way, things began to go wrong. The telephone hookup to the computer went down. Terminals were overheating in the sun. These kinds of problems are not forgiven in tournaments, when time limits are strict.

But the machine won its first game. Then its second. And its third.

By its fourth hands-down win, the tournament was in an uproar. Outraged human players actually attempted a mid-tournament coup d'état and physically seized the announcer's microphone. The Revolt of the Humans, it was called. The computer had no business being brought in against human players who were so much weaker, some players said. Those who never had to play against the computer had a huge advantage, they complained.

Eventually order was restored. The program won five games and lost none, outclassing all 128 humans in the Masson tournament. It was the first such win for any computer, and the first major test outside computer circles of what is now called the brute-force approach.

"We were as surprised as anyone else," Slate says. "It turned the consciousness around on what computers could do." From that time the use of brute force in chess computing seemed to dominate the field.

Ironically, it was believed for centuries that what made a grand master a great chess player was a prodigious memory and the ability to see 20 moves ahead and look quickly at hundreds of various possibilities. This is what the computers now do; it turns out to be exactly the opposite of what great human players do.

Grand masters, and other good chess players for that matter, actually see ahead on any one move only about 6 half-moves, not 20. A half-move is one player's move; frequently there are many reasonable responses to a single half-move. As for computers, Ken Thompson and Joe Condon's Belle now searches eight or more half-moves deep and examines about 50 million moves altogether on its turn.

The startling fact is that grand masters do not search much but actually manage to see only the very best moves. They do not even glance at other options.

This ability does not come from an innately powerful memory. When A. D. deGroot scattered chess pieces randomly on the squares and asked grand masters to look for five seconds and then reconstruct the array, they could not do it. They managed to recall where only three or four pieces were. Chess novices did exactly as well. But when the pieces were arranged on the squares as they might be in a real game, suddenly the grand masters and masters could recall the placement of virtually all the pieces correctly. Those ranked expert also did well, recalling about 72 percent of the pieces. Lower-ranked players recalled about 50 percent. Novices could recall only about 33 percent.

The grand masters recognize chess positions without having to examine them piece by piece. They see configurations on the board as a reader sees a phrase in a book, as a single unit of meaning. With each chess position, or "phrase," no doubt, the grand master automatically associates some moves, the best moves, and ignores legions of others.

This power simply to "see" situations and best moves makes it possible for a chess master to play simultaneously against 30 or 40 opponents and still win all but one or two of the games.

The chess master's ability to see in this way develops over decades of practice at the game. Putting such chess knowledge into computers has been difficult. All researchers in the vanguard of artificial intelligence face the same challenge, in a sense to give computers experience and knowledge of the world.

The problem has been that applying the sort of rule-of-thumb playing knowledge used by humans —and there are probably hundreds of these rules in chess—takes up great amounts of computer calculating time. In a tournament chess match there are only about three minutes per move. Thus, programs that are very smart don't have time to look at many possible moves.

In the end, programs that emphasize chess knowledge always face the most embarrassing type of loss: the loss by blunder. Robert Hyatt, the builder of the Cray-Blitz program, which took second place in the National Computer Chess Championship last year, at one time ran a chess program that emphasized knowledge over speed searches.

It was disheartening, he says now, because there are hundreds or thousands of bits of chess knowledge to be put into a machine, and then as many exceptions and special cases as there are rules.

Because there are gaps in the knowledge that can be put in, "in the course of a game, the computer is bound to make a mistake, miss an important move. Sometimes the error is not enough to lose the game," he says, "but too often it is." It puts the programmer on the edge of his chair throughout a match, wondering, *Will it see this? Will it miss that?*

David Wilkins, of SRI International, a nonprofit research and consulting organization, worked on knowledge-based chess play when he was a graduate student at Stanford University. "It is easy to put in knowledge, but the problem is that if there is even one piece of knowledge missing, the program will make catastrophic errors." For example, he says, a terribly embarrassing two-move checkmate can result if a programmer fails to include some knowledge of back-rank checkmates—those in which the king is sitting behind three pawns and can be mated by putting a rook on the back rank with the king.

The stories of mistakes made from chess ignorance are voluminous. One describes a program that had an unstoppable win but failed to see that it could just push its pawn across the board to be queened.

Hans Berliner, a pioneer in artificial intelligence and particularly computer chess, says that even the brute-force programs can be erratic. "They really have weaknesses in their understanding. I've seen Belle try to capture a pawn, at a cost of completely dislocating its pieces across the board, and then lose."

Thompson says that Belle sometimes strays into awful situations. "The computer will generate weaknesses in the opponent's position, but then it won't pick on these weaknesses. Just their existence is what it wanted to see. So then it just sits there and loves its position."

Some of the pioneers of artificial intelligence 25 years ago would have been amazed to hear of such embarrassments. Early successes led to rash predictions. Working on the assumption that a small group of powerful problem-solving techniques was behind all human mental abilities, researchers created programs like the Logic Theory Machine.

The program was given a few mathematical abilities and was asked to derive the basic equations of logic, an assignment sometimes given to college freshmen. The program quickly found proofs for 38 of the 52 theorems it tried in *Principia Mathematica,* the epic work of mathematical logic by Alfred North Whitehead and Bertrand Russell. The program also found one proof even more elegant than the one proposed in the book; this feat is still talked about a quarter-century later.

At the same time the Logic Theory Machine made its impressive debut, Arthur Samuel, then at IBM, built a checkers-playing program that could beat its programmer and even learn a little from games as it played them. It once beat the state champion of Connecticut, who was also one of the best players in the nation.

There followed more successes: a program that solved geometry problems, one that did word problems in algebra, and one that accurately imitated the choice-making ability of the investment officer of a bank.

Typical of the excitement of those times is this quotation from Herbert Simon, now a Nobel laureate: "It is not my aim to surprise or shock you, but the simplest way I can summarize is to say that there are now in the world machines that think, that learn, that create. Moreover, their ability to do these things is going to increase rapidly until—in a visible future—the range of problems they can handle will be coextensive with the range to which the human mind has been applied." In 1957 he predicted that in ten years "a digital computer will be the world's chess champion."

Then further advances failed to occur. The researchers ran up against the problem they still face. Yes, cleverly organized calculation can put on good performances, surpassing many humans. But no general intelligence, or even a limited one that has some real judgment, is possible without broad and deep knowledge of the world or some area of it. The human mind does not use sophisticated calculation; it uses instead a vast quantity of stored knowledge to effect judgment.

When programmers began to study what they

call knowledge representation, artificial intelligence entered a new phase. By knowledge representation they mean how information in huge quantities—such as the grammar, semantics, and vocabulary needed for language—can be stored so that it can be called up instantly and rummaged through, as we do in thinking.

From the newer, "knowledge-based" programming in artificial intelligence have come such things as the "expert" programs like those that now routinely do things once thought to require the subtle judgment of the human mind. One built at Stanford University, called Mycin, can diagnose some types of infections and recommend treatment. Another example is Dendral, a program that can determine the structure of organic molecules from spectrometric data. Prospector, at SRI International, is a program containing the collective wisdom of "a number of world-class geologists," according to Peter Hart, director of artificial intelligence at SRI.

One of those who has moved rapidly within the field as it expanded from its first embarrassing days is Herbert Simon. He is now an expert on the mental processes of experts. "Experts solve complex problems considerably faster and more accurately than novices do," he wrote in an article in *Science*. "Those differences are commonplace, yet only recently have we begun to understand what the expert does differently from the novice to account for this superiority." The differences, it turns out, are much like those between a grand master and a novice: The expert is able to see problems and solutions quickly, based on a combination of perception and deep knowledge and experience.

Simon and his associates have concentrated their machine intelligence work in chess, algebra, and physics. But they point out that what they have found can be applied to other areas of knowledge. "Chess, algebra, and physics are serving as the *Drosophila, Neurospora,* and *Escherichia coli* [the key organisms researchers have long used to study problems of biology] of research on human cognitive skill."

Some researchers are trying to make the very successful brute-force programs in chess also capable of using chess knowledge. What brute-force programs don't know is astounding. "A lot of what people assume has been in chess programs for years and years is not, in fact in there at all," says Slate. As an example, even low-class human players know that when the game gets down to a king on one side and a king and knight on the other, the game is a draw. Many chess programs don't know this and will play on hopelessly.

This is the case throughout the field of artificial intelligence. Programs that seem smooth and smart in one brief patch of running can be ridiculous in another. One famous example is from Joseph Weizenbaum's computer program called Doctor. It mimics a psychotherapist using the techniques of Carl Rogers:

(The program speaks in capitals, the patient in lower-case.)

Men are all alike.

IN WHAT WAY

They're always bugging us about something or other.

CAN YOU THINK OF A SPECIFIC EXAMPLE.

Well, my boyfriend made me come here.

YOUR BOYFRIEND MADE YOU COME HERE

He says I'm depressed much of the time.

I'M SORRY TO HEAR YOU'RE DEPRESSED.

It's true. I'm unhappy.

DO YOU THINK COMING HERE WILL HELP YOU . . .

This sounds good, but try an equally representative snatch from the time that journalist R.D. Rosen talked to the Doctor about his "menu anxiety":

Sometimes I'll be eating at an Italian restaurant, I cannot decide whether to order fettuccine or clams casino, and I become very agitated and disoriented.

HAVE YOU TRIED

Tried what?

TELL ME ABOUT WHAT

What the hell are you talking about?

WHAT THE HELL ARE I TALKING ABOUT

Can you help me overcome my anxiety concerning menus?

WHY DO YOU SAY YOUR ANXIETY CONCERNING MENUS . . .

In the realm of chess, some similar computer disorientation is clear. It comes naturally from using a mere brain to play chess, rather than a mind, which would comprehend chess more deeply.

Slate's program, called Nuchess (for Northwestern University Chess), is rated unofficially in the expert category at about a 2,100-point level (Points are won or lost in tournament play. Experts hold between 2,000 and 2,199 points. Masters hold 2,200 and up.) "But," Slate says, "my program is like a D-class, thirteen-hundred player having a very good day. The world's luckiest and strongest thirteen-hundred player, because it doesn't know a thing, but still it finds nice tactical combinations."

Since a computer in operation is no more than a great cascade of arithmetic in a box, Slate says that a computer playing chess "wanders through a space of positions that it doesn't understand. It wanders into something where it turns out that it is winning. It seems [in the view of the computer] to happen almost by chance. It suddenly says, 'Aha! I seem to be winning!'

"The computer doesn't know what it's doing from move to move, but it finds clever ways of surviving, and it wears out its human opponent," he says.

This "spacey" sort of chess can actually create a psychological advantage for machines against human opponents. "Our program has won against people by getting into ludicrous positions where its king wanders all over the board or something," unnerving the human player by dragging him into unchesslike situations.

But at the heart of the power of brute-force

chess is the simple fact that examining many possible moves—thousands of totally worthless ones as well as one or two brilliant ones—gives the computer a chance to find the best move. In simplified form, here's how brute-force searching works: The computer considers making one move, and its opponent's reply, followed by another move by the computer and so on for eight or so of these half-moves. The machine considers every possible alternative in a rapidly branching tree of moves and then assigns a number to the board position at the end of each tree branch. This value depends on how many pieces the computer would have, and other factors. At the end of each search, the basic brute-force chess-playing device moves in pursuit of the highest payoff. The whole process requires no sophistication.

Many of those who have built computer chess programs are now trying to add more chess knowledge to their programs.

"I have to get some sort of notion of attacking weak points, putting the pressure on," Thompson says.

Slate adds, "The key is to have a smart, brute-force program. What I want to do is put in enough chess knowledge to make Nuchess a sixteen-hundred player having a very good day. A sixteen-hundred player on a good day is probably a master."

One thing Slate has added to his program recently is a bit of knowledge for the end game, the stage of play following a major reduction in the number of pieces: that the game can be won by a pawn and a king facing only a king. Computers tend to push the pawn up first, but that leads to a draw. Winning requires using the king to out-maneuver the opposing king. Today Nuchess not only can win such maneuvering but will look ahead and trade off pieces to get to the two-on-one situation.

Wilkins, in his work at Stanford, developed and proved a method by which the computer can use concepts such as how to capitalize on a weak defensive array around the king.

Berliner, at Carnegie-Mellon University, has built a backgammon program that two years ago beat the world champion—the first time that a computer program had beaten a world champion at any board or card game. But what was additionally unusual about the match was that the program won, not by brute-force calculation, but by more subtle, humanlike play.

Berliner built into that program—he believes it can be a powerful technique in chess as well—a kind of master control system. This new function (called SNAC in the backgammon program, for *s*moothness, *n*onlinearity, and *a*pplications, *c*oefficients) gave his program a real feel for the game and where it stood; SNAC permitted choice in the style of play—more aggressive or passive—as the game changed. It allowed the program to make use of a whole array of new knowledge.

"It seems SNAC functions are the proper means of capturing the characteristic that human beings call judgment," Berliner says. "They make it possible to respond to small changes in stimuli with small changes in behavior, and this is exactly what judgment (as opposed to deduction) is all about. The next five years should see remarkable advances in computer game-playing."

By most estimates, a computer will beat the human world champion in 20 years. There is already enough powerful hardware, and plenty of ideas on how to put more chess knowledge into machines, to ensure the computer's victory.

The remaining problem is that the entire development of computer chess has been done by researchers sneaking an hour here and there in spare time. Government funds simply are not available. The political cost would be too high if someone found out that tax money was being spent on games.

Never mind that computer chess research, even without funds, has produced a number of programming advances that may be useful in such "expert knowledge" programs as the ones that help a doctor diagnose disease. There are no boundaries between disciplines in artificial intelligence. From tricks to make programs run a little faster to fundamentally important research in the understanding of human thinking, it is one field, populated by some Renaissance men. Simon, for example, one of the pioneers in artificial intelligence, won the Nobel Prize for Economics in 1978.

Computer chess programmers speak wistfully of what might happen if they had more time to spend at their machines. Fred Schwartz, who runs Chaos, a chess-playing program at the University of Michigan, says, "We have about twenty-five man-years' work outlined [to put more knowledge into Chaos], and about one month to actually work on it in the next year.

"What if Ken [Thompson] spent a full year on building up his evaluation function? I think he would make another big leap forward. Maybe not all the way to grand master, but pretty far," Schwartz estimates.

He concludes with a story about the last World Computer Chess Championship, held in Linz, Austria, in 1980. At the end of regular play, Belle and Chaos had tied. Belle with its brute-force approach looked at something like 150,000 positions in a second during the match. Chaos, emphasizing a far more subtle evaluation of moves, looked at only 85 positions a second.

Before the sudden-death match to decide the winner, Schwartz recalls a conversation with Thompson: "In a moment of exuberance, Ken said, 'Well, I'm only doing a hundred fifty thousand nodes a second, but I've got a plan to do two hundred fifty thousand nodes, and I'll be back next time with that!' " Schwartz, equally high on the moment, replied, "Great, Ken, but we'll come back next time with one that searches only ten nodes a second!"

Belle won the close contest. But no one is writing off the future of subtle machines, or discounting the finesse of the human minds that design them.

PART FOUR:
HUMANWARE

WIZARDS
OF SILICON
VALLEY

By Gene Bylinsky and Zhenya Lane

Silicon Valley is not some barren lunar crater or black crevice in the ocean's depths. Rather it is a lush triangle of unusual real estate stretching 30 miles to the south of San Francisco, along placid San Francisco Bay, to the Santa Cruz Mountains. In this verdant place where prune orchards and wildflowers blossom even in February, something else has burst into full bloom: 1,000 innovative science and high-technology companies flying flags like Hewlett-Packard, Intel, Syntex, Varian, Atari, Andros, and Zoecon. From the first seedlings of this new empire—planted barely a century ago—Silicon Valley has become the world's leading center of industrial innovation. Silicon Valley also mass-produces millionaires. *Fortune* magazine estimates that no fewer than 500 high-technology millionaires live there, many of them still in their late twenties and early thirties. No comparable mecca of high tech exists anywhere else. The closest counterpart is Boston's Route 128, that golden crescent of high-technology firms hugging the outer reaches of Beantown. But Silicon Valley long ago surpassed Route 128 in both the number of companies and the diversity of their products.

Birthplace of electronic games and home computers, of those tiny computers-on-a-chip known as microcomputers and of the world's most powerful supercomputers, of cordless telephones and digital thermometers, of laser technology and computer memories, of food colors and additives ingeniously designed to be harmless to the body—Silicon Valley is all that with much more to come.

Surprises like an artificial heart that requires no potentially poisonous nuclear fuel, bacteria engineered to make human insulin, computers that understand human speech and talk back—all are under intensive development in Silicon Valley's sparkling laboratories.

In some ways Silicon Valley is like medieval Spain, a launchpad for great expeditions into new worlds. But no territory in the world's history has launched more far-reaching, heavily financed journeys to unknown places than this tiny valley at the edge of the Pacific.

"The effect on Earth of Silicon Valley will be as dramatic over the next two centuries," says resident and computer manufacturer John Peers, "as the effect that [Dr. Louis Leakey's discoveries in] the Rift Valley will have on the evolution of man."

The Santa Clara Valley—to give Silicon Valley its proper name—is located on San Francisco Peninsula. It extends as far south as San Jose, the newest California metropolis, which to the surprise of many non-westerners is already bigger than Pittsburgh or Minneapolis. At the northern tip of the peninsula are all the attractions of that jewel of cities, San Francisco, and across the scenic bay are the distant lights of Berkeley and Oakland.

TRACKS OF THE FLY

No description of Silicon Valley would be complete, however, without mentioning Palo Alto, cradle of the first budding technologies in the area. Palo Alto is split right down the middle by El Camino Real, the broad highway that runs much of the length of California. Stanford University and Stanford Industrial Park, along with such various other citadels of science and technology as the Stanford Linear Accelerator and Linus Pauling's Institute of Science and Medicine, lie to the west of El Camino.

To the east is the city of Palo Alto itself. And it was right here, as you can read on a plaque outside a white clapboard house at 913 Emerson Street, that the marching tramp of a common housefly ushered in the electronic age and paved the way for the wizards of Silicon Valley.

It was a dramatic moment on that memorable

day in 1912 when a group of excited young men leaned over a table to watch a housefly saunter across a sheet of drawing paper. The fly's footsteps were amplified by a vacuum tube, making them sound like the steps of a marching soldier. This was the first application of the vacuum tube as a sound amplifier and generator of electromagnetic waves. The tube's inventor was Silicon Valley's first true giant: Lee DeForest.

The vacuum tube made possible such electronic wonders as radio, television, the long-distance telephone, electronic computers, tape recorders, and electronic eyes that open doors in stores and office buildings.

DeForest and his associates were then working for the Federal Telegraph Co., in Palo Alto, the oldest American radio company. But development of electronics in the San Francisco Bay Area dates back even earlier, to the turn of the century. Talented young men living in the area propelled the growth of radio by building the first major wireless station and by establishing the first wireless contact from an airplane to the ground. Federal Telegraph became the dominant force in this nascent industry.

The company proved to be the nursery of the first generation of the valley's wizards, for among the many bright young men it attracted in addition to DeForest were such men as Charles Litton, who later founded the giant Litton Industries, starting it in a garage in San Carlos.

An event almost as dramatic as that fly's monstrous march was the invention of the loudspeaker by two former employees of Federal Telegraph, Peter Jensen and E. S. Pridham. One day in 1917 the two men set up their apparatus on Mare Island in San Francisco Bay. From a window the loudspeaker faced the dock in the city of Vallejo, about a quarter of a mile away. The town's streets were deserted, but there was a man on the dock. Jensen's voice boomed over the loudspeaker, asking the startled man to remove his hat. He promptly did, apparently thinking he had heard a voice from heaven. That year Jensen and Pridham established the Magnavox Co., which manufactures loudspeakers and radios.

Federal Telegraph continued to breed other giants. While working at the company, Frederick Kolster developed the radio detection finder. In 1921 Ralph Heintz founded Heintz and Kauffmann. This company devised and built advanced shortwave radio transmitters, including those used by Rear Admiral Richard E. Byrd in his South Pole explorations.

The man most responsible for the snowballing buildup of new high-technology companies in and around Palo Alto before World War II, however, was Frederick Terman. The son of the developer of the famous Stanford-Binet intelligence quotient (I.Q.) test, Terman studied as an undergraduate at Stanford and took a doctorate in electrical engineering from MIT. In 1925 he began teaching a course in radio engineering at Stanford and soon started the university's radio communications laboratory. He attracted gifted students, and the fame of the laboratory spread. But it bothered

Terman that the scarcity of local jobs forced most of his graduates to go into "exile in the East," and so he began to encourage them to start companies near the university.

The biggest payoff came in 1937 when two of his brightest students, William Hewlett and David Packard, started a company on a part-time basis in the one-car garage of the house where Packard and his wife lived. The two young inventors began by making an audio-oscillator, a device that generates signals of varying frequencies. Terman recalls that he could always tell when the fledgling firm had received an order. "If the car was in the garage, there was no backlog. But if the car was parked in the driveway, business was good."

That garage shop is known today as the Hewlett-Packard Co., the world's largest producer of electronic measuring devices and equipment. The company now employs more than 42,000 people worldwide, including some 12,000 in Silicon Valley. The company's annual sales are approaching $2 billion.

Many other famous companies came out of Stanford University. In 1937 Professor William H. Hansen teamed up with Sigurd and Russell Varian, young brothers and backyard inventors in Palo Alto, to develop the klystron tube. A variant of the vacuum tube, the klystron generates strong microwaves that can be focused like the beam of a searchlight. The klystron tube became a foundation of radar and microwave communications, and out of it grew Varian Associates, a lucrative and prestigious company.

During World War II Terman headed a big defense electronics project at Harvard, where, among other things, he developed the aluminum chaff, which Allied bombers dropped on Germany and Nazi-occupied countries to confuse the Germans' radar. When Terman returned to Stanford, he continued to fan innovative flames. In another pioneering move, for instance, he set up Stanford Industrial Park near the university which became the prototype of such facilities. It induced still more companies to locate in the area.

Although it may seem as if Terman built Silicon Valley singlehandedly, there were other influences on the area's growth. William Shockley, coinventor of the transistor, for instance, returned to Palo Alto, his boyhood town, in 1955 and set up Shockley Transistor Corp. The transistor, of course, was the successor to the vacuum tube, perfected in Palo Alto 40 years earlier, and Santa Clara Valley was the logical place to cash in on this electronic technology.

A brilliant scientist, Shockley gathered around him a large group of gifted electronics specialists whom he had picked from big companies and universities throughout the country. In 1959, however, his operation fell apart as those bright young men, led by Robert N. Noyce, then only thirty-two, left and, with the backing of Fairchild Camera and Instrument Corp., founded Fairchild Semiconductor in Mountain View, near Palo Alto. While there, Noyce became the coinventor of the integrated circuit, the successor of the transistor, which now jams thousands of microminiaturized transistors onto a tiny chip of silicon. He also built up Fairchild Semiconductor into a $150-million-a-year operation. He left in 1968 and with his friend Gordon E. Moore, a talented chemist who had contributed some of the major advances in semiconductor technology, founded Intel Corp., in Santa Clara. About 14,000 people are now employed at Intel, which expects to have sales of about $500 million this year.

MICROBOOM

Intel became the brightest star in the hottest high technology going: semiconductors. The company pioneered a computer memory chip that has become an industry standard, and more recently it has introduced that revolutionary microcomputer —a computer-on-a-chip, which has led to the creation of many new consumer and industrial products. This pioneering, in turn, has contributed to the emergence of still other new companies, which are incorporating the tiny electronic devices into new consumer products.

One of the microcomputers' most spectacular applications has been the creation of electronic games. Nolan Bushnell, an engineer who went to Silicon Valley after having worked his way through the University of Utah by operating a game arcade in a local amusement park, was largely responsible for the birth of electronic games.

Bushnell began in a proverbial garage. (The process of small-company formation in Silicon Valley, incidentally, has been honed to the point where in some industrial parks budding entrepreneurs can rent a garage, complete with a roll-up door, and two or three offices adjoining it.) Later Bushnell moved the company he named Atari into a medium-sized one-story building alongside an apple orchard. Inside this building long-haired kids assembled games to the sound of rock music. More recently Atari has moved into huge quarters in nearby Sunnyvale. A cavernous game room off the main lobby is usually filled with excited youngsters playing fabulous electronic gadgets for free. For the most part, they are employees' children celebrating their birthdays.

The remarkable growth of Silicon Valley companies is a wonder to behold. One year you may visit a company founder in crammed quarters shared with a handful of fellow dreamers. Next year you may be visiting him in a spacious factory, which turns out data disks, or whatever he makes, like so many McDonald's hamburgers.

That kind of growth is what has made employment in Silicon Valley expand at seven times the national rate during the past five years and almost twice as fast as elsewhere in California. Jobs go begging for both specialists and the unskilled. This year an estimated 19,000 jobs will be available in Silicon Valley.

The beautiful setting and attractive job market have drawn many newcomers to the affluent communities of the valley. Real-estate prices have soared, the housing is now in very high demand. Many workers have begun to commute to the area from the outskirts, making automobile-generated pollution an increasing problem. The cost of living is high, too.

Yet most people already in Silicon Valley would not exchange it for any other place on Earth—so enamored are they of the climate and their surroundings, which includes a friendliness and informality not usually encountered in the big cities on the East Coast.

BECKONING MECCA

The valley is also changing in subtler ways. It is, for example, becoming more a professional, and less a manufacturing, center. Now the young fortune seekers are colonizing such obscure places as Aloha, Michigan, and Nampa, Idaho, where they are putting up plants because land and labor are cheaper. It has gotten harder to become a millionaire in the valley, partly because of higher taxes and restrictive federal regulations. However, new companies are continually being formed in the valley, and young men continue to get rich. Spreading applications of microcomputers, in particular, have recently given rise to a whole battery of companies that manufacture home computers—among them Apple Computer, Inc., Video-Brain Computer Co., and Cromemco—as well as chains of computer stores, such as Computerland and ByteShops.

Entrepreneurs now arrive from faraway places to establish companies in the valley. John Peers came all the way from England because he felt that Silicon Valley offered the best expertise for manufacturing his unusual product—a talking computer called Adam. For similar reasons, David and Doris Bossen moved to the valley from Columbus, Ohio, and started Measurex Corp., a highly successful company that makes computer-guided controls for paper mills and other manufacturing plants. As they explain, "Paper mills are in the woods because that's where their raw materials are. We are here because our raw materials are brains." The Bossens knew that the types of diverse specialists they needed could be found only in Silicon Valley, and they found them easily.

There is a lot more company development to come. According to Bob Noyce, semiconductor wizard and cofounder of Intel, the applications of microelectronics have yet to create a change as fundamental in our society as the automobile did. But he predicts that they will create such a change in applications where "slices" of electronic brainpower will be incorporated into a myriad of products for the home, office, and factory—from the telephone to the computer-controlled lathe. The recent appearance of those ubiquitous electronic wristwatches, pocket calculators, and electronic cash registers is just the first swelling of the ocean of products roaring up on those slices of electronic intelligence created by Noyce and Moore.

As for semiconductor devices themselves, Noyce adds, the technical problems have largely been overcome. Innovation in the semiconductor area, he thinks, is mostly over—at least for the time being.

Maybe so. But to find out for sure, we had to check with the financial backers of these contemporary Merlins who have the ability to transform equations into LED wristwatches, desktop computers, and bacteria that breed human insulin.

In their suite atop the Embarcadero Center, which houses their operation—with sunlit panoramas of San Francisco hills and billowing sails on the bay—neither Gene Kleiner nor Tom Perkins seemed much alarmed about any decline in innovation in the valley.

With good reason. Venture capitalists Kleiner and Perkins—whose previous successes include Fairchild Semiconductor, which Kleiner helped start, and a laser company that Perkins founded and sold to Spectra-Physics, the major laser-producing company in the world—are more active than ever with new and successful companies. Tandem Computers, specializing in multiprocessor "fail-proof" computers, was one of the few companies able to go public in 1977, a tough year for such enterprises. Another of their new brainchildren, Genentech, a firm working in recombinant DNA, has already successfully engineered bacteria into microscopic factories that churn out human insulin.

"Another one of our companies," Kleiner says, "is developing an artificial heart." That company is Andros, in Berkeley.

"Although we're looking at many different companies, and helping to develop some here," Perkins says, "I don't think we'd dream of financing a new semiconductor company." The costs of doing that have soared into tens of millions of dollars.

Kleiner and Perkins sometimes lend money to new businesses and leave them to their own resources, but they often take a more direct interest in new companies. Both Tandem and Genentech, for example, are being run by people who worked for Kleiner and Perkins in those same Embarcadero Center offices before setting out to chart new seas.

Well, then, if not semiconductors, what do these two ambitious capitalists think the wave of the future is going to be?

"If you look across the horizon," Perkins says, "we think the next wave is biological."

BIOMED WHIZ KIDS

Two other giants of the valley, Alejandro (Alex) Zaffaroni and Carl Djerassi, actually got this biological revolution going. The smooth-talking Zaffaroni was born in Montevideo, Uruguay, the son of a banker. He started out by studying medicine, but, as so often happens, a brilliant instructor soon redirected his interest into biochemistry. The instructor had explained in exciting terms the central role of the carbon atom in organic chemistry, and Zaffaroni decided to explore that role. He came to the University of Rochester and obtained a doctorate in biochemistry there. Soon his brilliant flashes of insight produced what is known in chemistry textbooks as the Zaffaroni System, a method for separating steroid compounds by paper chromatography. This method served as a stepping-stone toward large-scale production of steroids by pharmaceutical companies. Several years later Zaffaroni became executive vice-president of Syntex Corp., in Mexico City, where he led the company's pioneering drive toward the synthesis of the birth-control pill and other advanced drugs.

At this stage the ubiquitous Fred Terman enters the picture once more. In his effort to build up Stanford University's chemistry department, Terman, as the university's vice-president, asked Djerassi to become a professor there. Djerassi did so—without leaving Syntex. Djerassi is the father of the birth-control pill, which he developed while working for Syntex. He would be a giant anywhere. Born in Vienna, Austria, of a Bulgarian father and an Austrian mother, both physicians, he was expected to follow in their footsteps. Like Zaffaroni, however, Djerassi was drawn into chemistry by an outstanding instructor, receiving his Ph. D. from the University of Wisconsin in 1945. Medicine's loss has been chemistry's gain; according to a friend who is a Nobel laureate, Djerassi has done enough high-quality work to win two or three Nobel prizes.

Both highly creative and imaginative individuals, Zaffaroni and Djerassi have since been responsible for the formation of four pioneering companies, all located in Palo Alto. To accommodate Djerassi, in a modern mountain-comes-to-Muhammad move, Syntex relocated its entire research and its manufacturing operation to Palo Alto—thus bringing still another high-technology company into the area. Zaffaroni came from Mexico to head the Syntex operation. While both men were with Syntex, they originated Syva Corp., which jointly with Varian Associates engaged in the manufacture of medical instrumentation, and Zoecon Corp., a firm pioneering the applications of hormonal regulators of insect growth. Djerassi later left Syntex to direct Zoecon, while continuing to teach and direct research at Stanford.

In 1968, after leaving Syntex, Zaffaroni started Alza Corp., which is putting into practice his novel ideas about drug delivery. Zaffaroni has always felt that methods of dispensing medications have not progressed much since the time of the ancient Egyptians. So Zaffaroni set out with some novel ideas: delivering drugs through the skin via impregnated patches; developing a birth-control device that releases tiny amounts of progesterone inside the uterus for a whole year; devising drug reservoirs akin to microminiaturized spaceships that, after being swallowed like pills, become anchored inside the body to release finely controlled amounts of medication.

Alza has also produced an imaginative offshoot; Dynapol Corp., which is developing another of Zaffaroni's ideas—food additives and preservatives so structured that they harmlessly pass through the stomach without entering the bloodstream to cause possible damage. This is done by "leashing" the smaller additive and preservative molecules to harmless inert molecules that are too big to penetrate the walls of the stomach.

Meanwhile, Djerassi's Zoecon—from the Greek word zoe ("life") and *con* for *control*—began to explore the fascinating idea of synthetically imitating the growth-regulating hormones that are naturally present in insects. The idea was to use these as novel insecticides—harmless to humans and other vertebrates but deadly to specific insects. Zoecon has created a number of successful products, already being marketed, including an insect-growth regulator that keeps mosquitoes and flies from maturing into destructive adults from their harmless larval stages. The same product, applied differently, prolongs the life of the silkworm, thereby inducing it to produce higher yields of silk. This was Zoecon's first product to be introduced in Japan. Another of the company's projects is aimed at developing insect pheromones, or sex attractants, as possible lures for use on sticky traps.

Federal regulations have been a barrier to more rapid progress in wider applications of hormonal insecticides. Instead of encouraging the American farmer to use something as cleverly contrived and as safe as synthetic copies of the insects' own hormones, government regulatory agencies have been erecting excessive and sometimes picayune obstacles in the way. Similarly, Alza has been delayed in the introduction of some of its devices. (Both Zoecon and Alza are now subsidiaries of larger companies. Alza is a subsidiary of Ciba-Geigy; Zoecon is part of Occidental Petroleum; Zaffaroni and Djerassi continue to head their respective companies.)

Not one to be deterred by small setbacks, Zaffaroni predicts a great future for the applications of the new biology. "It's difficult right now to see the future applications, because we are at the beginning of the recombinant-DNA technology," he says. "It was difficult to map out the products created by the transistor. But the applications will be far, far wider.

"The first things people can see right now to do with this new technology are to produce agents used in pharmacology, therapy, vaccines—produce all the rare chemicals, that are in the body, develop plants with higher abilities to fix nitrogen or with higher protein levels. These are the simplest ideas. If you want to let yourself go further out, who is to say we can't construct biological memories—DNA memories? The way biological memories are constructed is most fantastic, in very small space."

But there is a problem. It's not easy to anchor the technology by patents, since the basic work is coming out mainly from the universities. This makes it hard for this new industry to develop a capital base.

A DELICATE BALANCE

And this is one of the paradoxes of the Valley of the Giants: The scientist-businessman, just as often as the academician-industrialist, must continually strive to balance pure research with practical realities. Djerassi probably has been the most successful high-wire walker. "I'm one of the few people at Stanford, maybe anywhere," he says, "where for twenty years I've led a completely bigamous life, all that time serving as a professor of chemistry with a very large research group, and not just teaching courses, and at the same time being either a vice-president at Syntex or a president here at Zoecon. I think it was worthwhile for both places. I became a much better academician and a much better government adviser by spending part of each day running a high-technology industrial enterprise. And I became a much better and more innovative corporate executive by being primarily a scientist, which is more difficult to learn than business. Business you can learn on the job. Science you cannot.''

It's unlikely that Djerassi could have led that kind of life anywhere else. Liberal-minded universities that encourage professors to participate in company building and to serve as consultants are the exception rather than the rule. Attempts abroad to duplicate the valley's remarkable company-formation process have scored only the most moderate successes. Undaunted, foreign investors have decided to settle for second best and have already begun pouring funds into Silicon Valley. Platoons of West Germans can be seen signing into local hotels on visits to semiconductor companies in which they now have substantial interests. The Japanese have gone further, setting up advanced companies staffed by American engineers and scientists. From these pioneering companies, the Japanese then pump the newest technology into their own burgeoning industries. This has led some inhabitants of the valley to fear that their peaceful abode will someday become a battleground between those two erstwhile World War II allies.

The failure of clone Silicon Valley abroad is not really surprising. If we look back on the startling developments that have sprouted from this fertile center, it would be a mistake to underestimate the creative influence and spirit of entrepreneurism that a few outstanding individuals—men like Terman, Shockley, and Djerassi—have generated in their wake. Their ability to transform new ideas into successful and innovative products undoubtedly has been a major force in attracting young talent to the valley, making it a remarkable breeding ground for new industries within less than a century's span. The history that they have helped to create and Silicon Valley's unique environment cannot instantly be duplicated.

Whatever the future holds, the technological revolution that flared up in Silicon Valley shows few signs of sputtering out. Judging from the myriad new companies started in the last few years, only the first harvest has been reaped from the valley. Time will tell, but one thing is certain: The footsteps of that Palo Alto fly already have resounded around the world many times over.

THE TINY WORLD OF
CLIVE SINCLAIR

By Tony Durham

Clive Sinclair's pocket-sized television sets use picture tubes squashed to the thinness of a magazine. His shrunken computers, like the one he's holding in the photo at right, fit into his palm. In the future Sinclair foresees robots the size of corpuscles, little electric cars, and missiles as small and slow as turtles. As Sinclair pursues his quest to build a Lilliputian world of electronic products, he himself has become a modern Lemuel Gulliver, an adventurous explorer who appears to grow bigger all the time.

But, unlike Gulliver, he will not be tied down. He's about to turn loose on the video market a million pocket-sized TV sets a year. In the drive to come up with an inexpensive, mass-appeal Volkscomputer (a term co-opted by a competitor, Commodore Business Machines, for its $299.95 VIC 20), Sinclair has produced what one Commodore executive admiringly calls "the motorbike of computers," the Sinclair ZX81. Prices of these diminutive products match their scale. His computer costs $149.50, his TV about $100, both low enough to attract consumer attention and to make Sinclair richer than he already is.

The inventor of the "motorbike" computer lives a Mercedes life. With his wife, Ann, teen-aged daughter, and two sons, he shares a large stone house on the edge of green fields in Cambridge, England. The ground floor is a series of echoing, white-walled spaces linked by monastic stone archways, with views onto a peaceful, tree-edged lawn. He reads poetry and sometimes science fiction, through thick glasses. His idea of relaxation is doing a mathematical puzzle in *Scientific American*.

But Sinclair, lean and energetic at forty-one, with hair the color of spicy ginger, is not one to sit still for long. He seldom watches his little TV sets, or any television. He has run in two New York marathons, in 1980 and 1981, finishing each time in a little more than four hours. He relishes the company of fellow members of Mensa, most of them on the trail of intellectual pursuits. (Is Mensa stuffy? "It strikes me as elitist," Sinclair replies, "but not unpleasant.") Sinclair's business rode a roller coaster in the 1970s, profits plummeting after appalling production snags with a new digital watch. Government investment kept the firm from derailing completely, and finally Sinclair set off on his own track in 1979. He left the prosaic but profitable instrument-making side of the business behind. But he hung on to his research team and his pet projects.

Now he's roared back up: Sinclair Research sells more personal computers in the United Kingdom than any other manufacturer, and it ranks third in international production of personal computers, behind Radio Shack and Apple.

Sinclair has almost always earned his living by thinking small. At the age of twelve, in a family of engineers, he was inventing mechanical calculators in the school workshops. His interest shifted during his teen years to electronics, television, and computers. After their marriage, Clive and Ann built up a small business, selling transistors by mail order in their spare time. From components they moved on to kits.

In 1962 you could buy any number of kits for transistor radios. The first Sinclair Radionics kits were distinguished by their minuscule size. Perhaps out of sheer curiosity, people bought a radio half as big as a matchbox, even though there was no speaker and though it had to be used with an earphone. Schemes for a pocket TV set were already in Sinclair's mind, but he knew it was not yet remotely feasible. Instead, he produced salable but technically unadventurous hi-fi kits, with a look that has remained the mark of a Sinclair product: compact, stylish, and unfussy, with a penchant for mat black rather than gleaming aluminum.

Then, early in the 1970s, Sinclair took the first of several business risks, tying his fortunes to pocket-sized products. Most manufacturers thought a pocket calculator was impractical. The power demands were too heavy for small batteries, they argued.

Sinclair disagreed. With his team he devised a circuit that slashed the calculator chip's power consumption to less than one tenth of what it had been. Hundreds of times a second the circuit switched the chip off, then powered it up again before the data faded from its memory. "Texas Instruments was very surprised, but it certainly worked very well," he says.

In all fairness, it ought to be mentioned that there is some question about how surprised Texas Instruments (TI) was. A spokesman for TI says Sinclair did not invent the "power-down mode" feature, nor is it used in all hand-held calculators today. Furthermore, TI says with a touch of grumpiness, the hand-held calculator was invented in the mid-1960s by Texas Instruments. The first successful commercial model was introduced by Canon, with technical assistance from TI, in 1970. Still, in a show of sportsmanship, the spokesman calls Sinclair "a highly respected innovator."

Sinclair is prouder still of his second calculator, the Cambridge Scientific. Once again the little Cambridge firm took a commercially available four-function calculator chip and tortured it into a performance far beyond its original specification. Sinclair had to devise new ways of calculating the logarithmic and trigonometric functions before his colleague Nigel Searle could reprogram the commercial chip. The methods then known to mathematicians would have generated more numbers than the chip's tiny memory could store.

Doing more with less became a habit. Like a custom-car builder, Searle once more took a standard chip and turned it into something more powerful. This time the result was a machine that you could teach to do long calculations at the touch of a single key—not the first programmable calculator, but the first to use just one chip.

Sinclair had pioneered small, cheap number crunchers, but soon the Americans and the Japanese were beating him at his own game. His next project, a digital watch, misfired badly, and Sinclair Radionics virtually bled to death. It was saved by massive transfusions of government money, and by the modest success of the Microvision, a book-sized television set with a 2" screen. This was, as Sinclair admits, "something you would use occasionally rather than a lot," and it cost more than twice what a normal black-and-white portable was selling for. But about 2,500 people a month decided they would like to own one. It prepared the market for the subsequent wave of tiny televisions made by the Japanese.

Few guessed at the time, but the Microvision was for Sinclair only a stepping-stone toward his long-term goal of a flat-screen television. Unfortunately for him, his government shareholders were unwilling to take the next step. So he had to take it on his own. He left the company he had built, sold his house and his Rolls-Royce, and started from scratch with a new company called Sinclair Research.

The new firm's first product—a stripped-down computer—was a winner. At the time, Apple, Commodore, and Radio Shack had gone after the small-business market. Texas Instruments, Atari, and Mattel were trying to seduce consumers with sound and color, fun and games. Sinclair aimed below them all with a machine that looked more like a calculator than a computer. He brushed aside existing standards in such matters as interfaces and programming languages and built his ZX80 as if it were the first computer in the world. For most of his customers, of course, it was.

Not that it was perfect. Its 21 chips, Sinclair says, did the work of 40 chips in other systems. "The penalty," he admits, "was that the screen flickered when you entered data. But that seemed a slight price to pay." So did the price: $200.

The irritating flicker has disappeared from the new model, the ZX81. And now there are only four chips in the whole machine. For $149.50, it comes with 1K of memory (it holds 8,192 bits of information). That's a modest capacity (for $150 more, the Commodore VIC 20 has a 5K memory), but it's sufficient for simple home budgeting tasks, math lessons for children, or space-war games without a lot of fancy graphics. Currently the ZX81 cannot be hooked up to information systems such as The Source (which provides news and other data through home telephones feeding home computers). Nor does it have the capacity to be the centerpiece of a word processor.

Computer buffs are enthusiastic about one ZX81 feature in particular: one-key entry of most commands. To get the computer to print something, for example, users simply type *P*. Most other machines won't respond unless operators spell out the command: P-R-I-N-T. Overall, says Taylor Pohlman,

174

of Apple, Sinclair has probably boosted the sales and the profits of his competitors.

"He's introduced a lot of people to the idea of using a computer," Pohlman says. "His computers are cute. And after people operate them for a while and find out that computers aren't threatening, they'll move up." Although the move to an Apple can be quite a leap—the least expensive system runs about $1,300—the Sinclair ZX81, Pohlman says, is likely to be good for Apple's business.

So far it's been good for Sinclair's business, too. In the first seven months after its introduction in the United Kingdom, Sinclair sold 100,000 of his little computers. Recently he clinched a distribution deal with Mitsui to sell them in Japan. In the United States he broke new ground by persuading American Express to include the ZX81 among the goods offered to card-holders in the firm's frequent mail-shots.

Profits from this international enterprise are now pouring into one of Sinclair's long-cherished projects: a television set only slightly bigger than a deck of cards. He and his design team have developed— *perfected* was his word—a small, slab-shaped TV tube in which electrons start at the edge and travel sideways across the screen. The Timex factory in Dundee, Scotland, is getting ready to produce a million sets a year for Sinclair. In the immediate future the screen will be merged with a Sinclair computer to make a self-contained system, possibly with projection equipment to enlarge the image.

Further in the future Sinclair foresees still greater opportunities for trimming the size of machines, from cars to guided missiles.

He says that the time and the technology are already ripe for an electric city car, with a range of about 70 miles per charge. With computer programs, his design staff is investigating possible body shapes and studying how batteries would stand up to differing patterns of use. Almost everything about the proposed Sinclair car is either secret or genuinely undecided: so it is hard to see just how it will differ from the electric cars that appeared sporadically, and flopped commercially, during the 1970s. But Sinclair talks conspiratorially of new materials, new types of motor and control systems, and a dashboard liberally bedecked with computers.

The car won't rust, he says, and will always start. People will come to prefer it to their long-distance cars, and then "they might say to themselves. 'I'd like to use this all the time, and I'll rent a big car when I need one.'"

Can Sinclair's logic end man's romance with the motorcar as easily as that? "I think that will pass, just as the size of the television screen has ceased to be a status symbol." Perhaps the new status symbol will be a car that you can fold up and put in your pocket.

Sinclair expects that many other break-throughs will come about with help from his little friends: the computers he and his competitors have sent out from their factories, covert agents of cognition. What may come marching back, he says, is an army of intelligent robots.

"I think that when you put computers out among so many people, as we and other people in the personal computer business are doing, and you bring up a generation that isn't frightened of the present generation of computers, you will see things change dramatically," he says. "Somebody's got to come up with that breakthrough that enables us to make apparently intelligent machines, and you just need a lot of people thinking about it for that to happen, I suspect."

It is a fascinating theory: that the massed Apples, Pets, and ZX80s of the world, combined with the fertile minds of their users, will provide the prebiotic soup from which the next product of evolution, the Intelligent Machine, will emerge. Sinclair does not put it in quite such strong terms. But suggest to him that computer hobbyists have produced nothing more useful than 20,000 versions of Space Invaders and he leaps to their defense.

Doctors and lawyers, he says, have written serious programs for use in their own professions. They find it easier to do it themselves than to use a programmer. A schoolboy recently wrote a chess program for a ZX80. Amateurs are transforming the way computers are used. It won't be long before someone—possibly an amateur—delivers a robotic housekeeper.

Sinclair would welcome it at the threshold. "I'd like to have robots in the home," he says. "Nice little servants. Chop wood and things." Foot in the door, the robots would move on to other chores. Some would travel inside human arteries and veins, broadcasting the view by television, performing operations with assistance from a surgeon outside. That scenario, Sinclair thinks, is possible in fewer than 20 years. "I suppose I could do it in ten," he says, "if there was a lot of money."

One evolutionary step further, robots will replace dogs as pets. The human master will learn to love them. "I think that's likely to happen." Sinclair says. "Exciting. And then they'll get brighter than us. And they'll take over." There will be no place for humans in a society of intelligent robots, he says, except perhaps as pets.

It won't happen in his lifetime, he believes. That leaves plenty of time for generations of miniature robots to wait on us—and do some of our dirty work.

Flying machines the size of bees, for example, could buzz over our cities, looking for muggings, robbery, traffic accidents, mayhem, and then summon help. Much more controversial, courts could order prisoners to be fitted with electronic bugs, forcing felons to live under 24-hour surveillance. The proposal angers some civil libertarians, but Sinclair happily grasps the nettle: "Give criminals the choice," he says, "I bet they don't choose to sit behind bars."

Other robots could wage war, using Sinclair's version of a missile shell game: the exploding tortoise. Sinclair sees it as a logical step following the contraction of the ICBM into the cruise missile. "You want to go a step further," he says. "You want a tortoise. It will make its way slowly but surely to where you want it to go and then blow itself and the surroundings to bits. You'd make them by the million. Set off a million at once. They'd never stop them."

War and crime are the only small clouds Sinclair sees in a bright future of helpful, portable machines. His vision is all the more credible because of what he says he *can't* do. Sinclair has often dreamed of a convenient personal flying machine, something as easy to use as a car. After careful study, he's concluded the blue-sky idea will never get off the ground. You'd better believe him.

COMPUTER KIDS

By Doug Garr

Enter a typical, middle-class teen-ager's bedroom with the obligatory rock and roll posters, in this case Led Zeppelin and the Kinks, adorning the wall. Greg Trautman, fifteen, with thick eye-glasses and braces on his teeth, is a precocious-looking talkative New York adolescent. "Still programming in BASIC?" he playfully asks his friend Sujee DeSilva, who is busy ordering his Atari 800 to cough up the modest game program he wrote. "We've had the Atari for only about two months. So we're just getting into the intricacies of it," Greg says, as if the new machine were partly his own. Trautman is clearly envious of his friend's recent acquisition, mainly because he must be content with his year-old Radio Shack TRS-80, a less costly home computer with a much smaller memory. The Atari has four times the memory, high-resolution color graphics, and a greater potential for complex programs. Jason Bucky, fifteen, is jealous, too. He also has a TRS-80, but he's pining for a disk drive, a device that would let him eliminate the slower, more cumbersome method of storing his programs on cassette tapes.

Later Carey David, also fifteen, arrives. Greg acts positively reverential. "He's the genius," Greg remarks. Carey looks like a street-smart kid right off the set of *Welcome Back, Kotter,* but a few moments after he slips his cytogenetics program into the Atari, it is apparent that he is serious about computers.

A bunch of jumbled letters appear on the video display terminal, which Carey says represent amino acids, or chains of protein molecules. "A geneticist can use this program to manipulate the DNA chain in order to change the molecules and get different organisms," Carey says. "Like insulin." To demonstrate, he types a little more on the Atari keyboard. In 1981, while a ninth-grader, Carey took five months to write this program. It won first place in the Queens Borough Science Fair. His current project is a program that will let his junior high school work out the complex administrative tangle of class scheduling.

The interests of this small group of computer devotees are typical of what has happened with computer technology among the young. Ten years ago the only contact the average person might have had with a computer was some basic math done on a pocket calculator or a few games of Pong in a bar. Today the computer has entered the mainstream of American life as a serious and powerful learning tool. And nowhere is the impact of this change more evident than among the school-age kids of today, the first true computer generation.

Already the computer plays a crucial role in the lives of thousands of kids, both as the center of an all-consuming hobby for youngsters like Greg Trautman and his friends and as a classroom fixture. According to the National Center for Education Statistics, by 1980 52,000 computers were already in use in the nation's elementary and secondary schools, and by the end of 1981 an estimated 40,000 more machines were installed. Even some recreational activities have become computer-centered. A slew of summer computer camps, where kids do everything from playing softball to programming in BASIC (a primary computer language), are now operating across the country from Moodus, Connecticut, to Zaca Lake, California.

All indications are that the proliferation of machines is only just beginning. The Tandy Corporation, which owns Radio Shack, has already awarded $500,000 in computers and equipment to schools all over the country under its Educational Grants Program. Since the fall of 1979 the Apple computer company has donated nearly $1 million worth of equipment to schools under a similar program. Even the federal government is beginning to see the value of computer education. As the result of an Apple Corporation offer to donate one computer to every elementary, junior high, and high school in the country—a total of about 80,000 computers—a group of California congressmen introduced the Technology Education Act, which would give high-tech corporations larger tax credits for such contributions. The National Science Foundation has also been asking computer firms to donate equipment to high schools under its Development in Science Education Program. And to train students to use these machines, the Educational Testing Service, of Princeton, New Jersey (the organization that develops and administers the Scholastic Aptitude Test), recently announced that by 1984 it will have standardized its advanced curriculum in computer science.

In the meantime schools are not waiting for the trainers and equipment to arrive. They have already begun to revamp their curricula at an astonishing rate, so much so that by the 1990s it will be all but impossible for a student to pass through high school without having had some first-hand experience in using a computer. These changes are not just happening in well-to-do private schools, where this might be expected, but in public schools, too.

A good example is Francis Lewis High School, part of the New York City public school system, in Queens. Because of the foresight of its principal, Melvin Serisky, it is already serving as the model for the rest of the system. When the school opened in 1960, Serisky was a math teacher. His involvement with computers began in 1961, when a bank donated to the school an old Burroughs LGP-21 machine, roughly twice the size of his desk. Mostly it was used to perform some of the more mundane administrative tasks: drawing up absentee lists and keeping track of report cards. Once, Serisky recalls with some satisfaction, he also used it to catch a graffiti artist known only by his spray-painted moniker, "Fred 169." Serisky simply asked the computer to call up all the students with the first name Fred and all students with 169 somewhere in their street address. The list produced six suspects. From there, catching up with the perpetrator was an easy matter.

Since then Serisky has used the machines for far more libertarian work. In 1964 he managed to procure for the school a new Olivetti computer. By then he began to worry about the possibility that the computers would become elitist tools. Traditionally, the first children who got to use them were the gifted math students. Other students should learn how to program computers, too, Serisky thought. So even though Francis Lewis is situated in a predominantly white, middle-class neighborhood, Serisky made an effort to recruit nonwhite students. "As a coach goes after football players," he explains. "I went out to get black kids for the computer courses. The first two minority kids who came here are now in science—at Harvard."

Today it is impossible to get a diploma from Lewis without some exposure to the terminal, much the way it is impossible to graduate from other high schools without English composition. Computer languages like BASIC, FORTRAN, and COBOL are all taught at Lewis. It is assumed that every student has at least a minimal amount of computer literacy so that when one needs some information about a college, he is expected not to bother the guidance counselor but retrieve the data himself from the computer.

Much of the education with computers at

Francis Lewis still emphasizes math heavily—a student can get the equivalent of ten years of math training if he wants—but computers are used in social studies, the sciences, and even the business department. In social studies, for example, students get a civics lesson in guns and butter when they play a program known as HMRABI (for Hammurabi), adapted from a games program written at the Digital Equipment Corporation. In HMRABI the student becomes the leader of ancient Babylon and, through the mediation of the computer, conducts trade in arms and distributes wheat to his minions. He remains in power as long as he is popular, that is, as long as his decisions on trade and arms produce a strong military, adequate food supplies, and a stable economy. One wrong move, however, and there is a peasant revolt, ending in the leader's going into exile.

Educators everywhere are initially inclined to reserve this kind of computer education only for gifted teen-agers, but this bias is changing. Thanks to pioneering experimenters like Seymour Papert, a professor of mathematics and education at Massachusetts Institute of Technology, we are beginning to shed the misconception that computers are only for older kids who are "good in math." Papert, an innovator in education, studied with the noted Swiss child psychologist Jean Piaget, and it was he who developed an ingeniously simple language called LOGO that lets kids begin programming computers virtually before they can read.

As a result, the machines are becoming more accessible to other age groups. For example, elsewhere in Queens, Halsey Junior High School has a very progressive curriculum in which thirteen- and fourteen-year-olds regularly demonstrate their skills at school science fairs. At one recent show, Sergio Rico, fourteen, displayed an elaborate program he had written, explaining the earth's hydrological cycle. It was complete with graphics and a multiple-choice quiz. Another student, Sujee Danesh, only thirteen, prepared a display depicting blood flow through the human heart.

The architect of the Halsey computer curriculum that produced these kids is a young, energetic ex-engineer named Howard Weinman, who finds teaching much more rewarding. "Unlike us, these kids are growing up with the machines. It will be natural for ideas to come from their work with computers. Next year," he says, beaming, "we start kids programming in BASIC. I'm extremely envious I didn't have these opportunities."

Children even younger than those of junior-high age are getting, or will soon get, these same opportunities. At the Lamplighter School, a private institution in Dallas, children ranging in age from three to nine are learning LOGO on a battery of computers as part of an experimental program, a collaborative effort of the school, the Texas Instruments Corporation, and MIT's artificial-intelligence laboratory. And by the start of the academic year next fall the New York Academy of Sciences will begin an experiment in New York in which children in first grade and kindergarten will be exposed to computers.

As educational tools, the computer and the LOGO language have also demonstrated spectacular potential among autistic children and those with

187

learning disabilities. For example, at MIT a team of educators reports working with a thirteen-year-old whom others had written off as retarded. Today he is programming computers with the skills of a professional.

And though it may seem farfetched, the computer has even had positive effects on so-called bad kids. In Downey, California, Larry Bauder, a computer-store owner, has been working with tough kids from the slums of Watts and East Los Angeles since October 1981 under the auspices of a group called the Los Angeles Committee for the Transformation of Street Gangs. Every Saturday, from 11:30 A.M. to 3 P.M. about a dozen kids from the ages of ten to seventeen spend their time sharpening their skills on Commodore computers (on loan from the company) that Bauder reserves for them in his store.

"Even after twenty years as a teacher," Bauder says, "I'm still amazed at what happens to the so-called dummies when they sit down in front of a computer. For one thing," he notes, "they learn how to read. Some are functionally illiterate, but if they want to learn how to play Star Trek, they learn how to read."

He has been optimistic ever since he tutored his first group. After only four months of lessons, Bauder claims that four inner-city girls he trained are fully qualified to run a computerized office. "They really take to the computer," he says of all the kids he has taught. "I have to kick them out of here at three o'clock." Bauder now hopes the new step will be to move computers into Watts and East Los Angeles.

One reason for the universal appeal of the machines among the young is that the kids have a much healthier attitude about the computer than many technophobic adults do. "My parents see the computer as something that's going to take over the world. I see it as a way of expressing myself," says Jason Bucky.

Though children who visit the Marin Computer Center, in Corte Madera, California, take to the machines eagerly, the staff there discovered that some sort of acculturation course was necessary for adults to overcome their biases. "They think the machines are just waiting to embarrass them," says director Mary Cron of parents who visit the center.

Two educators, David and Annie Fox, founded the Marin Center in September 1977 with their own money. Its basic premise is that children should be exposed to the wonders of computing. Cron says she herself is the least likely person to guide such an institution. She admits she was skeptical of the computer as a teacher, and thought it would merely spew out prerecorded information, with no opportunity for student-teacher interaction. Now she views the machines as important additions to the classroom. "A computer can't replace a teacher, but it can free him or her to do more quality work," Cron says.

The Marin Center's basic programming teaches youngsters "sequential" thinking, computerese for simple logic. More significant, the computer has inspired dramatic improvements in the performance of many so-called slow learners. Many of the children who need remedial instruction begin computer courses with the idea that the machine is smarter than they are. But the quick discovery that it's really a stupid device that needs to be told what to do makes them comfortable with it.

So far the kids who have dominated the headlines are the wizards who seem to have been born in front of a video display terminal, like Stephen Bacchus, who at the age of thirteen was the youngest freshman ever admitted to New York University. Bacchus is heavily into advanced calculus, but he's also done programming on the past performances of racehorses. One program he wrote for his Commodore Pet microcomputer has come up with a number of winners, and Bacchus is proud that his machine's handicapping ability is decidedly more advanced than his father's. Now Stephen is building a robot.

Eugene Volokh, who has a fair amount of entrepreneurial skill, started his own software firm, VE Soft, and has worked as a consultant with Hewlett-Packard for several years. He's written a program for the company's HP-3000 computer that now sells for $1,500. "Eugene knows more about the HP-3000 than just about anyone else," says Clifford Lazar, Hewlett-Packard's manager of systems development and sciences. Much of this Eugene did while working on an undergraduate degree in mathematics and computer science at UCLA. He is now fifteen years old.

These are the stars, the superachievers of tomorrow. It is likely that they would be precocious even without computers. But in the future thousands, perhaps millions, of average youngsters will be using the machines in more routine ways. And they will not have to worry about becoming computer-adaptive the way most adults do right now.

Steven Senzig, an independent programmer and computer-store owner, puts it this way: "The kids who are now four, five, eight, ten years old will learn how to program just as you and I use the telephone. They're going to be downright scary to old types like us who weren't raised that way." As a result, when Senzig opened a computer store in Lansing, Michigan, he wasn't surprised to find as his first customer an eleven-year-old who had diligently saved the money for a computer from his earnings as a newspaper delivery boy.

There is also a dark side to this fascination of the young with computers. Many kids have cajoled their parents into spending their hard-earned savings on a microcomputer, the price of which can run from a few hundred dollars to well into the thousands. Even then many of these kids often cannot afford to buy the expensive software (the programs, usually recorded on tape cassettes or magnetic discs, needed to run the computers).

Rather than write their own programs, kids will often pirate others—despite the elaborate protection codes that have been put into them—in disregard of copyright laws. Right now a program for microcomputers known as Locksmith is a best-selling item among the young computer users. Locksmith amounts to a sort of skeleton key that can break through the protection codes of a great many software programs. The kids in effect use it to copy programs illegally, much the same way you might use your cassette recorder to tape a friend's record album.

Greg Trautman sees some personal irony in software theft. "Ninety percent of the programs used by my friends have been copied, but I hope this changes. Stealing programs adds to the high cost of software. And," he adds, "I'd like to write and sell software."

As computer usage has become more pervasive, so, too, has widespread "crashing" (sabotaging) of central data banks. Sometimes it is done out of pure avarice; sometimes it is merely a prank. A few school kids in California tapped on to their school's computer records of its grading system and programmed it to give themselves A's. They gave everyone else F's. In a more malicious vein, a couple of years ago two Chicago area teen-agers broke into DePaul University's main computer system via a computer-telephone hookup and shut it down during registration week. Authorities didn't realize their machine had been sabotaged until someone noticed a threatening message that appeared on a computer terminal: "If you don't give us a mixed-assembly software program, we'll shut you down again."

That same year some industrious youngsters from the Dalton School, an exclusive private school in New York City, patched into the huge electronic files of a data-systems firm in Canada. The culprits—four thirteen-year-olds—managed to erase one fifth of the computer's memory before they were apprehended. The case became a cause célèbre among security experts and computerphiles because of the high degree of sophistication involved in doing this.

It may be several years before we can assess the

full impact of computers on society, because children have only just begun using them. "Computers are still new, and there's a lot of confusion as to what they are able to do," explains Ann Lewin, director of the Capitol Hill Children's Museum, in Washington, D.C., which has 30 Atari microcomputers in its Future Center. "It will take about a decade to tell what kind of effect the use of computers will have on education."

But already the effect of computer technology on the young is the focus of controversy among adults. Many conservative communities are complaining that kids are becoming "addicted" to video games and spending most of their time and all their lunch money on them. This is probably an overreaction. Bright kids like Trautman, Carey David, and Sujee Danesh do spend after-school hours at the local pizza parlor playing Tempest and Centipedes, this year's replacements for Pac-Man and Space Invaders.

The games are entertaining and provide a way to escape from the pressure of growing up. But the kids understand that computers are merely computers. They have learned the programs, they respect the work that went into formulating these programs. Children are doing just fine in putting the computer in their own perspective. Jason Bucky says, "If I'm feeling bad, I'll go home, switch on my TRS-80, and kill a few Xylons."

"I've been amused that towns are trying to outlaw video-game arcades," Ann Lewin says. "I'd rather see them playing games than doing drugs."

Trautman, the whiz from Queens who's

desperately trying to trade his Radio Shack TRS-80 for an Atari 800, makes another observation: "Have you ever been to the Penn Station arcade at three o'clock in the afternoon? It's all businessmen in suits playing video games." Then he adds, "Well, I guess it's better than having them hang out in bars."

VIDEO WIZARDS

By Phil Wiswell

Bernie DeKoven is a little round elf of a man with a bald head and bushy beard who likes to spend some of his more creative moments sitting on his sun-drenched porch in Palo Alto, California, thinking such thoughts as *What would I really like to play with?* His thoughts are of special interest to those in the billion-dollar video-game industry since what DeKoven likes is very often what several million people will pay to play.

DeKoven is a senior game designer with a Silicon Valley computer software company named Epyx Automated Simulations. His job is to create new games that will be developed and sold for use with home computers like the Apple II and the Atari 800. DeKoven's success at this can be measured equally by his comfortable salary and his happiness with the medium of the microcomputer. At forty years of age, he has been a game designer—both as a freelancer and as an employee of companies like

Ideal Toys—most of his adult life. In all those years, he admits, he has never found a tool for creating games as exciting and as versatile as the computer. "It's the Swiss Army knife of the mind," he says.

Right now DeKoven is unique in the burgeoning video-game industry. One would expect him to have an engineering or computer science degree, but he has neither. His academic background consists of an undergraduate degree in English literature and a graduate degree in theater. He doesn't write computer programs and makes no pretense about being an expert on the technology. The skill he brings to his work is strictly the creativity of a game designer.

Because of the technical nature of the medium, video-game manufacturers have usually relied on computer programmers to do their designing. And it shows: If you run your finger through the list of game categories today—shoot-em-ups, race games, and so on—you will find at most one or two original ideas in each group.

Some of this situation is changing. As more younger players find it increasingly easier to beat, or "max out on," current games like Space Invaders and Pac-Man, and as more adults beyond the average twelve- to eighteen-year-old age group become fascinated by video games, the manufacturers are beginning to see a demand for games that are more intellectually sophisticated and more challenging. And so they are turning to more creative types, game designers or computer programmers who have the originality of a good game designer.

What new games are being programmed now? To answer that question, I took a walk through the game-design labs of three of the most creative software companies in Silicon Valley to get a sneak preview of what's under development. But before talking about that, a little background would be helpful.

Most video-game players begin at the pay-for-play arcade games. In time they reach the point where they desire electronic entertainment on their own terms—having the game available day or night, with no long lines and no obligation to pay 25 cents each time. That is why *home* video-game sales soared to $1.2 billion in 1981 and are expected to double this year.

Today home video games (as opposed to arcade games) can be divided into two categories: ROM (Read-Only Memory) cartridges used with popular game systems like Atari's VCS (Video Computer System) and Mattel's Intellivision, and cassette tapes or floppy discs for home computers like the Atari 400 and 800, the Apple II, or Radio Shack's TRS-80.

In the industry, Atari is the leader. It is the only company that makes games for arcades, home computers, and home game systems. This wide experience, along with its early entry of the VCS, has allowed it to get and hold 70 percent of the home video market. Today there are about 7 million of its VCSs attached to American television sets. After Atari comes Intellivision, which puts out a more sophisticated, and more expensive, system, in use in 1 million homes. Until recently North American Phillip's Odyssey[2] and Astrocade's

Astrocade were the only other home systems available, but this year they face competition from as many as seven new home game machines. Some, like Coleco's Colecovision and the new 5200 machine from Atari, will compete for a share of the market with more technologically advanced hardware. Others, from companies like Tryom and Emerson Radio, will compete by offering comparable machines that are priced lower.

But the majority of those rushing into the business are not building game-playing hardware. They prefer to concentrate on program cartridges that play on another company's system. The reason: Last year home video-game players bought 30 million game cartridges, paying between $20 and $40 for each of them. The big bucks are clearly in software.

Today a score of companies are making games for the VCS. Many of them are communication giants like CBS, Lucasfilm, 20th Century-Fox, Paramount, Thorn/EMI, and Disney Productions, all of which have recently formed video-game divisions. What no doubt inspired them was the fact that in 1981 Atari, itself owned by another giant, Warner Communications, brought in more money than the blockbuster film *Star Wars*.

Any video-game system is really nothing more than a limited microcomputer, and the microcomputer industry is still in its formative stage, marked by some growing pains. For one thing there is no standardization. Companies must now produce games to be played on specific machines. A game designed for the VCS will not work on Intellivision, nor will it work on Atari's home computers.

Similarly the Apple II, the Atari 800, and the TRS-80 computers cannot use one another's software.

Eventually, analysts predict, the video-game market for the VCS and microcomputer will merge into one. Using the attraction and familiarity of video games as a sales tool, the industry will mass-market home computers. There are about 10 million game systems in use in homes now and only about 1 million home computers, but by 1990 this ratio is expected to be reversed. For that to happen, manufacturers will have to produce a computer that offers three features: high resolution, or the ability to "paint pretty pictures" on the screen; high-speed graphics, the ability to move many objects around the screen quickly; and a low price tag. The best guess is this machine is at least three or four years into the future.

Currently the software firms are designing for both game systems and computers. One firm that has had phenomenal success at this in its brief three-year history is Activision. Last year, for example, its software sales totaled $65 million, which placed it second only to Atari. For the most part Activision's games have had a reputation for being graphically beautiful and easy to learn. Some, like designer David Crane's Freeway—in which the player tries to move his video chicken back and forth across a crowded freeway without having it get run over—are also humorous, even cute. In Pitfall, Crane's new game, the player must run his little man through a series of jungle scenes in which he searches for treasure, has to jump over cobras, scorpions, and rolling logs, and swings on a vine over tar pits, lakes that appear and suddenly disappear, and crocodile-infested swamps. Even Johnny Weissmuller would find it challenging.

Activision was the first company to jump on the VCS bandwagon. It was also the first to promote its game designers as superstars, according them almost the same status as performing artists. Activision game packages always include a photograph of the designer, together with liner notes from him offering tips on playing the game. As a result, the designers now receive an average of 7,000 fan letters and hundreds of phone calls *each week*. Some of the more prolific designers, such as Crane, often find themselves stopped on the street by twelve- and thirteen-year-old video-game fans and asked for autographs.

But Activision designers do not slavishly produce games with their fans in mind. "Our guys really design for themselves," says Tom Lopez, vice-president of editorial development for the company. "And they like to play games. I think that's most important. When I look for a game designer, the first question I ask is, 'What is your favorite video game?' I then ask for their best score. And there can't be any hesitation in their answers. They have to be video-game fanatics. They have to love it, live it, breathe it. Our guys design all day long, and for relaxation they go home and play games."

This, as I found out, is no idle boast. Activision's designers' idea of treating me to lunch was to go for a couple of slices of pizza, a cola, and several dozen video games at a Chuck E. Cheese Pizza Time Theater, one of a chain of pizza parlor-game rooms founded by Nolan Bushnell, the man who created Atari. The designers often go there to check out the work of the competition, to get new game ideas, and just to have fun.

At this stage of the business Activision still feels video-game design is more of a science than an art. Crane thinks a nontechnical person could not program successfully for the VCS because he or she would probably have to shelve 999 of 1,000 game ideas partway through the design process. Because of their limitations, the VCS and, to a lesser extent, Intellivision demand a high level of technical knowledge in a designer, Crane says. The VCS represents only about $20 worth of six-year-old electronics, and designers must know how to stay within the peripheries of this technology. For all practical purposes, *designer* and *programmer* are interchangeable terms at Activision.

The same holds for the creative group of Imagic, Inc., a smaller but no less eager software company that was formed a little over a year ago by two executives from Atari, two from Mattel, and half a dozen Atari game designers. Already they have developed seven VCS games and five more for Intellivision.

The development cycle for a new game at Imagic is four to six months, which might lead you to believe that a game designer does more playing around than is necessary. "The fact of the matter," says Dennis Koble, vice-president of software development and an Imagic designer himself, "is that every step of the process is so fraught with detail that I cannot write a section of code and expect that program to do what I envision."

The code Koble refers to is known as machine language, long strings of ones and zeros that form instructions for the microprocessor. The VCS and Intellivision don't understand English, or even BASIC, a commonly used, English-like computer language. They understand only machine language. To illustrate just how primitive the VCS is, one BASIC instruction to have the machine print a single word on screen translates to more than 50 steps in machine language. To help the programmer, designers use an "assembly language" that converts a program to machine code.

Essentially there are three steps that Koble, or any video-game designer, must go through when developing an idea. The first is to write a program in assembly code and store it. The second is to convert the program to machine language. And finally to try it out in the VCS or Intellivision and see whether it does exactly what it is supposed to do. "Since it never does," Koble says, "you go back and debug."

Because this process of debugging, or finding and correcting errors, can confuse even the best programmer, it is common practice to work on only one section of a game at a time. Typically, a designer will spend a week to more than a month just creating the basic game format, trying different colors and shapes to get the picture right. Particularly time-consuming was the right look for Atlantis, Koble's newest game, now under development for the VCS and Intellivision.

The idea for Atlantis had been on Koble's mind for months, but only as a very general concept. First he sketched out the idea on paper, keeping game elements at their simplest because, in his words, "a truism that has run through the industry for years is that we [the game manufacturers] have consistently overestimated the intelligence of the player and consistently underestimated his dexterity."

Atlantis is a game with a theme, a feature that players seem to enjoy. In Atlantis hostile aircraft besiege the underwater city, which the player must defend with three laser guns. Like many games of this genre, the player faces increasingly more formidable waves of attack that do not seem to underestimate his dexterity.

Still, there will be many players who will want more. And Imagic will give it to them soon with the release of a sequel to Atlantis, Cosmic Ark. When a player's last fortification in Atlantis has been destroyed, a small ship can be seen escaping from the city just before the game turns off. This intimates that while Atlantis had been annihilated, part of the player's force always survives and is able to get away. This ship reappears in Cosmic Ark, in which the player's goal is to repopulate Atlantis by flying to many different planets in the galaxy, fending off asteroid storms, and bringing two creatures (one male and presumably one female) back from each planet to the starship via tractor beams.

Like the most successful games, Atlantis and Cosmic Ark can be learned almost instantly, with one play of the game. "One fact that has been established by the coin-op industry over the years is that *nobody* reads instructions," Koble says. "I don't care who the player is, or what background or experience he has. *I* don't read instructions. If you can't figure out how to play, just inherently picking it up as you go, it isn't worth your money."

DeKoven agrees. In his view, the reason for this is that the video-game industry has unwittingly created a new and powerful language based on what DeKoven calls the "videoglyph," a combination of changing colors, shapes, and sounds linked to a physical response. For example, successfully moving a paddle to intercept and reverse the path of a ball on screen is a simple videoglyph. "It is a symbolic language that instantly conveys concepts, like the Egyptian hieroglyphs," he says. "Unlike hieroglyphs, however, the videoglyph is not static. It is a language whose characters can change as you read them." Part of the fun for players is that they are able to understand this language intuitively. A good video gamer can walk up to a new machine, plunk in a quarter, and stay alive, for a while anyway, without having read the instructions.

What DeKoven sees as part of his challenge of game design is to add to this vocabulary of video-glyphs. One thing that he thinks will expand it enormously is the home computer, which typically has a larger memory and is capable of more elaborate graphics than game-only machines. Games designed for it can therefore be more complex, move involving. Much of the computer game software available today falls into either of two categories: adventure games or arcade-style games, in part because programmer/designers have been content merely to use new effects with game concepts that have already been proven. "The microcomputer is a wonderful tool," DeKoven says. "Why use it to simulate only existing game concepts?"

Why, indeed, you may ask yourself once you've seen Ricochet, DeKoven's first computer board game. Not only is Ricochet a game of pure strategy (as opposed to hand/eye coordination), but a player's move is depicted on screen with arcade-quality graphics that make the game as entertaining to watch as it is exciting to play. And, unlike video games that are merely computerized versions of real-world games like football, Ping-Pong, or a simulation of war, Ricochet is something that could have been put together only in a computer. In real life, the game just could not be played without a frictionless billiards table.

Ricochet is a two-player board game with five variations. (For the sake of simplicity, I will describe only one.) Two humans can play against each other, or a human can play against one of four computer opponents, each with a different strategic style. A player occupies either end of the board and controls two video ball launchers and six pieces that look like bars that are initially arranged in a bowling-pin type of formation in front of a goal. With each turn a player has one of two options: shoot a ball from a launcher, or shift one, or all, of his pieces to block an opponent's shot or facilitate one of his own.

The object of Ricochet is simply to get the most points by the end of the game. The game ends when one player has run out of balls or when both his launches have been destroyed.

When a player fires a ball, it always ricochets off the sides of the screen like a high-speed billiard ball. Each time the ball caroms off one of the individual bars, it scores one point, and every time it ricochets off the opponent's goal, it scores ten points. The more ricochets—that is, the more tricky angle shots a person can make with each ball—the more points he can score. The essential strategy of Ricochet is deciding whether to make a shot or to move a piece. Like a pinball game, the more rebounds, the more points a player can get. Some players try an intuitive approach to keep the ball ricocheting, but more often than not it is a well-calculated shot that rewards the player with plenty of rebounds, plenty of points, and a nice graphics display. "This," DeKoven says, "is the end of my search for the ultimate bounce."

He says he can usually come up with a game concept in a week, as he did with Ricochet. Design and production of the game, however, can total eight or nine months. With Ricochet, he began roughing out the idea at his home studio, experimenting with boards of various sizes and shapes and using matchsticks to represent the elongated pieces. After a week of this, Jim Connelley, the president of Epyx, looked at the game design and was so fascinated, he sat down and programmed the video game—board and pieces—in one day so that DeKoven could go to work on the specifics. Originally, when the ball hit a goal, it was supposed to bounce out of play, but a bug in Connelley's program let the ball continue to bound into play and allowed the shot to last longer. DeKoven liked that so much, he incorporated this extended bounce into the game and named Connelley codesigner.

What makes Ricochet stand out from so many other games is that DeKoven created many variations for it, not just the one basic scenario most board games offer. This way a player can find one that suits his or her skill and interest levels. DeKoven now wants to reach further than that. "One of the delights of playing on a computer," he explains, "is that it allows you to enter other realities." Witness another DeKoven game, Alien Garden.

The idea of Alien Garden centers upon a figure-it-out-as-you-go experience that will subtly test and develop analytical skills. In an alien world presented to you by the computer, you move an insectlike figure through a videoscape in which there are 19 different kinds of crystals that do things like shrink, grow, explode, shy away, and so on. Some crystals are edible and give a player strength (and points). Some are deadly and explode on contact. And others will disappear from the scene if you stroke them with one of the creature's wings. Once someone has played the game long enough to perceive these laws, he will understand the behavior of the crystals. Like real crystals, the video ones will grow on the screen geometrically and according to their individual characteristics. Although it is supposed to be a game, it is just as pleasant to sit back and watch the crystals grow, the effects are so stunning.

Going from idea to machine code to any video-glyph is not simple even for the best of designers. DeKoven, who is happy with his job, admits that working on video games has its down side. What could be depressing about thinking up games? Well, for example, he has a concept for a game that would visually simulate the flight of an eagle, its hunt for prey, and the dive—all from the eagle's perspective. But he can't produce it, because it's impossible to produce the graphics that could portray the game realistically.

There are technologies that might help. One is to connect the computer to an interactive video disc containing thousands of detailed computer images. Using holographically portrayed surroundings in 3-D is another. Although both are feasible, they are also too expensive right now.

A Dekoven idea for the near future is his concept of a Fitness Arcade. It would eliminate the drudgery of keeping fit by giving the person exercising some sort of fantasy environment to work out in. Take something like a rowing machine. "The player slips a twenty-five-dollar token into a slot," DeKoven says. "In front of him is a very detailed video screen. He now finds himself sitting on a raft, which he must try to guide through whirls, eddies, and currents. He might even find himself being chased by frightening creatures."

The difficulty of the course shown on the screen and the tenacity of the creature would increase in proportion to how well the player is doing. The machine would also monitor the rower's pulse through the sensors in the handle grips. It would also monitor how many strokes the rower is making and how full they are. Each rower would have a maximum heart rate prescribed for him according to an exercise program, and the game would shut off once he reached that rate. The technology is already available to build this, DeKoven asserts. The cost: a few thousand dollars more than that of a Nautilus machine.

As futuristic as the Fitness Arcade sounds, DeKoven can top it. He recently visited a Berkeley, California, company named Autogenics, which makes biofeedback equipment for the medical industry. Autogenics has a machine that senses alpha waves, which are produced by the brain in a relaxed or meditative state, and that allows a person in an alpha state to move an object on a video screen simply by thinking about it. Instead of a joystick, the person uses a headband, and instead of snapping and jerking his wrist back and forth, the player just relaxes and tries to let it happen.

"The game that I am proposing for their machine," he explains, "is a telekinetic game. You would see on the screen before you a table with three fragile objects on it. As you meditate or relax, one of the objects starts to rise off the table. The better you get, the easier it is to elevate it and move it around. If you lose your concentration, the object falls and breaks and you lose points. As you begin to explore the possibilities, you will find you can make the objects move around simultaneously, perhaps almost juggling them. Then, of course, the game will try to keep up with your skills by introducing more objects onto the table. Did I tell you about my two-player telekinetic game concept . . .?"

PART FIVE:
COMPUTERS THAT WALK

ROBOTS THAT DUST

By Richard Wolkomir

Want a slave?

Think of it—a serf to vacuum your floors, cook duck a l'orange for you and your friends, serve, clear the table, and wash the dishes. After the party, your slave would help you relax with a game of backgammon. And, if your slave had the audacity to win, you could administer a punitive kick.

Disgusting? Abhorrent?

What if the price is right? What if everybody has one? And what if this slave is a thing of aluminum and transistors?

Eventually, we all will have robot slaves. So inexorable is the evolution of industrial robots, say the experts, that household spinoffs are virtually inevitable. In fact, they almost are here.

Joseph Engelberger, president of Unimation, Inc., the leading robot manufacturer, recently appeared on the Merv Griffin television talk show. With Engelberger was a Unimation robot, an intellectually feeble articulated arm. The robot stood in front of a conventional household window, complete with curtains. When Engelberger pressed a button, the robot opened the curtains, picked up a squeegee, washed the window, dried it, and put down the squeegee. Then it unlatched and opened the window, picked up a watering can, watered the petunias in their sill-top flower box, put down the watering can, and closed the window and curtains.

"The point," says Engelberger, "is that it's very difficult these days to find a human housekeeper who'll do windows."

Lest anyone think the president of Unimation is kidding, at one end of his office in a white colonial house in Danbury, Conn., he recently installed a kitchen. "Within the next few years, I will have a robot in a closet next to that kitchen," he says. The robot will be called "Isaac," in honor of Isaac Asimov. However, the robot will not write books. It will be a domestic thrall.

Isaac, an articulated arm on wheels, already is in training at Unimation's research facility in California. It can roll 40 feet across a floor, find an assigned spot, and carry out programmed tasks. In one test at Engelberger's office in Connecticut, *Isaac* opened a cabinet, pulled out a mug, poured coffee into it, and rang a bell to inform its boss that his coffee was ready. But Engelberger says that *Isaac* is still primitive, with much tinkering to be done before it takes its station in his office closet.

"In just three years, this will not necessarily be an entirely practical, useful device, but it will do enough to spark imagination," Engleberger says. "It will be under voice command, and it will take orders for coffee and Danish. It will be able to heat the Danish, get the cups and saucers, make the coffee, and serve it to my guests. And it will clean up afterwards and put the dishes in the dishwasher on command.''

No tipping, of course. No need for even a thank-you.

Isaac will not be the first domestic robot. An Urbana, Ohio, computer engineer named Charles Balmer already has built ''Avatar,'' which looks like the offspring of R2D2 and a dental chair, one of many home-mades now clanking about in U.S. houses. Atari-founder Nolan Bushnell's Androbot, Inc., is marketing two home robots. One is the Androbot, a radio-controlled automaton that dances and sings under control of a home computer. The second is BOB, a self-contained model with its own ''brain on-board.'' And the Heath Company recently introduced Hero I, a $2,495 robot ($1,500 for a kit) that carries up to a pound in its metal claw, ''sees'' (dimly), and performs such chores as delivering drinks and patrolling for burglars (see ''Heath's Hero,'' *Omni,* January 1983). Encountering an intruder, *Hero* will raise its gripper and announce that it is calling the police, which should be sufficiently startling to straighten the hair of most prowlers. On the other hand, Hero's repertoire of tricks is limited. In fact, one of the 33 phrases it intones is: ''I do not do windows!'' Heath expects Hero's chief use to be as a teaching device for robotics students.

''I don't think anything is now being done specifically to address the household robot market,'' says Joseph Engelberger. But Elliott Wilbur, an expert

215

on housing and a vice president at Arthur D. Little, Inc., the international consulting company, says that, although he cannot reveal the details, one of his firm's big-corporation clients is currently experimenting with domestic robots.

Wilbur was one of nine Arthur D. Little experts in fields ranging from electronics to home appliances who met recently to consider the prospects for domestic robots. In a dining room at the prestigious think tank's Cambridge, Mass., headquarters, over turkey salad, fruit, and sherry (served by humans), the consultants analyzed household robots as a potential product. They disagreed, often sharply, on exactly how robots would fit into the home appliances market. But they agreed on one important point: Developing the necessary technologies is not only feasible but virtually inevitable. As engineers steadily boost the IQ of industrial robots, they are creating the technological bits and pieces that eventually will feed your Siamese, rake your leaves, and knot your cravat.

It will not be a snap. Household robots must not crack the glasses or try to leave rooms through the wall, so they will require senses and abilities that today's industrial robots lack. To see how difficult developing those technologies will be, consider just one organ, the eye.

"Producing a digital image of merely a slice of bacon takes about 76,000 numbers," says Donald L. Sullivan, an Arthur D. Little, Inc., computer expert who is developing a robot inspector for industry. His machine looks at a product, such as a slice of bacon, sees the darker strips of meat and the lighter strips of fat, computes the fat percentage, and then either passes the slice along or tosses it in the reject bin. Sullivan's laboratory at Little headquarters is a creative jumble of video cameras, microprocessors, and—a Salvador Dali touch—slices of meat, pouches of sweet-and-sour pork, and crackers.

Robot vision, he explains, works by translating a video image into numerical values that a computer can understand. The system assigns a number to each shade of gray between black and white—the lighter the shade, the higher the number. To interpret a picture, it breaks down the image into dots, assigning an appropriate number to each dot. For the image of a bacon slice, for instance, the computer is programmed to identify all dots with values below a certain number as the dark background, above a certain number as white fat, and in between as red meat.

Compared to negotiating a house bump-free and cleaning the bathtub, it is simple for a robot to inspect bacon for fat, crackers for burns, or food pouches for leaky seals. Yet just developing one experimental inspection robot, Sullivan says, has cost about $70,000 in hardware and $300,000 in engineering time. He says that the far more sophisticated vision a household robot needs is four to nine years in the future.

Touch, too, is on its way. Researchers in Japan are developing hospital robots that can gently hoist a patient off his sickbed, deposit him in the bathtub, and return him to bed later. Within three years, Australian engineers expect to have robots sufficiently sensitive to shear sheep, with no bandaids needed.

Robots also must understand spoken commands. That technology is inevitable because it means a jumbo payoff in the office automation field. IBM researchers, among others, have already developed a typewriter that takes dictation. It is s-l-o-w, its computer requiring 100 minutes to transform into print a sentence it took only 30 seconds to speak. But IBM's engineers predict they will have a practical prototype in a few years. David Lee, an Arthur D. Little expert on consumer products and appliances, comments that "Voice recognition is well on its way, and that will be one thing that helps open the domestic robot market."

A household robot should speak, as well as hear. Already the "operator" giving users of New York City pay phones such messages as "Sixty cents, please," is a computer with a 70-word vocabulary. A decade or two hence, say robotics experts, when you tell your household "Isaac" to change the bulb in the bathroom lamp, it will answer.

Don't expect *Isaac* to run off to perform its humble chore. It is more apt to toll. A walking robot is possible; Robert B. McGhee, an Ohio State University electrical engineer, has built a six-legged walking machine. With sensors in each foot, this metal cucumber beetle can even pick its way along a path littered with stumps. And at Carnegie-Mellon University's Robotics Institute, visiting scientist Ivan Sutherland is developing a six-legged robot vehicle. Carnegie-Mellon professors have even choreographed a dance for a robot and a woman.

However, the first domestic robots are unlikely to lurch through your house on metal legs. "The wheel was one hell of a good invention," says Joseph

Engelberger. "There's a lot of fun in making walking machines, and they may be interesting for going over rough terrain. But a house is not rough terrain." He points out that Unimate's Isaac will roll through his office nicely on three wheels with tires that swivel, enabling the robot to move in any direction without turning its body. "It's as if a car could park by moving sideways," he says, adding that Isaac occupies no more floor space than Larry Csonka.

When will all these components come together? "My own conjecture," Engelberger says, "is that it will make economic sense for the luxury market by 1990." At their session on domestic robots, the Arthur D. Little consultants predicted that commercial models will make their debut a bit later, about the year 2000.

But the $64-billion question is this: Will anyone buy the things? As Arthur D. Little housing specialist Elliott Wilbur asks, "Why not just hire a kid to cut your grass?"

"Because there will be fewer kids to hire," responds Martin Ernst, a vice president at the consulting company and an operations research expert who believes dropping birthrates will reduce the supply of casual labor. He also believes that fewer adults will be interested in drudge work; on a recent trip to the Netherlands, he found that the entire country has only two commercial laundries because so few Dutch workers now are willing to take such jobs.

And according to one authority, we are already buying robots aplenty. Electronics engineer Stuart Lipoff told the Cambridge meeting: "We already have programmable microwave ovens, dish-washers, and swimming-pool cleaners. These are all robots of a sort, even if they don't have eyeballs.''

The consensus at Arthur D. Little was that robots will develop along two tracks. First, our appliances will get smarter and smarter. "The cost of braininess is so low,'' pointed out Martin Ernst, that eventually the smart appliances may begin to merge. "You might end up with a unit that combines a refrigerator and an oven,'' agreed appliances engineer David Lee. At a pre-set time, the freezer section would pop the dinner you punched in for tonight into the microwave section, which would turn itself on. The next step, according to Lee, will be an automated menu—choosing your dinners for the next month, perhaps—coupled with automatic inventory control based on supermarket package codes.

Donald Sullivan foresees a clutterless household in which the robot can store infrequently used gadgets in a central area. "One night you come home in the mood to whip up a gourmet meal,'' he says. "You tell your kitchen, 'Hey, forget that frozen dinner I programmed for tonight, and get me the Cuisinart.' It goes 'rumble, rumble, ptooey,' and out pops your Cuisinart.'' He even envisions a box of basic parts that the robot might draw upon to assemble household mechanisms as required.

On robot evolution's second track, engineers will be developing a stand-alone automaton to handle such odious chores as raking the lawn, cleaning the bathtub, and reading a toddler's favorite story over and over. Ultimately, the stand-alone robot and the smart appliances might share a central brain that controls everything. "You have the mechanical peripherals, with modest intelligence, and a basic computing engine that ties everything together," says Gordon Richardson, an Arthur D. Little robotics consultant. Someday your house itself may be a robot.

"Remember, the brain doesn't have to be resident in the robot; it might even be shared by all the houses on a block," Joseph Engelberger notes. He compares tomorrow's robot house to HAL, the invisible computer that controlled the space ship in *2001: A Space Odyssey.* But he says it also will be necessary to have a stand-alone robot, like C3PO in *Star Wars,* to handle chores like cleaning and to give people something to relate to. However, while the robot is stacking dishes in the kitchen dishwasher, its "brain" may be humming in the basement. And its "eyes" may be mounted in the ceiling of each room.

Eventually, says Engelberger, in its servant's quarters, the household robot will have spare parts for all the appliances in the house, with a collection of tapes giving maintenance instructions. "At night, you'll tell the robot, 'The range isn't working; please fix it by morning,' " he says. While you sleep, the robot is awake. It is alert for intruders and fires, of course, but it also is operating on your range. "If it gets stuck, it calls the factory and talks to a smarter robot to find out what to do," says Engelberger. If it lacks a part, it orders it. By the time robots have advanced to this stage, they also will keep your larder stocked, ordering replacement items from the supermarket as needed.

The repertoire of skills such a robot might master

seems unlimited. Arthur D. Little's Elliott Wilbur suggests that robots will be even more salable if they replace skilled workers as well as low-priced laborers. "For instance, your robot should be able to cut your hair any style you like," he suggests.

Whatever form the robot takes, safety must be engineered in: "You'd hate to have a two-thousand-pound robot go berserk in your living room," says robotics expert Gordon Richardson.

Household robots will have to be at least as safe as highly trained guard dogs, says Joseph Engelberger: "They might have a magnetic radiation aura around them, or some sort of sensor to detect a baby in their path." Domestic robots, he adds, will certainly include in their programming the Three Laws of Robotics, propounded decades ago by Isaac Asimov: A robot must not harm a human being, nor through inaction allow one to come to harm; a robot must always obey human beings, unless that is in conflict with the first law; a robot must protect itself from harm, unless that is in conflict with the first or second laws.

But will tomorrow's household slave actually be a machine? "The ultimate answer," says Elliott Wilbur, "may be to implant this intelligence in an animal, like a monkey." He suggests that a microprocessor collar might pulse out signals to guide the beast through a chore that require more-than-simian brainpower. Nor is the idea of zoological slaves farfetched, considering that for eons mankind has exploited the muscles and brains of beasts from llamas to sheep dogs. At the Arthur D. Little meeting, computer expert John Langley cited his milkman, who recently switched from a horse-drawn wagon to a truck and found the horse more efficient because it directed itself down the street and knew all the stops. "The milkman just ran along behind the cart, carrying the bottles to his customers' doors," Langley responded.

Researchers at the Tufts-New England Medical Center Hospital, in Boston, working under a National Science Foundation grant, are training monkeys to perform such services for paralyzed people as fetching food from the refrigerator, opening or locking a door with a key, removing a record from its album cover and placing it on the turntable, and brushing their owners' hair.

The first household robots may also debut as caretakers for invalids. Unimation, working with researchers at Stanford University, is attempting to modify a Puma robot to understand simple spoken commands to aid paraplegics.

Martin Ernst points out that as the population ages, such services will be in growing demand. And another Arthur D. Little engineer, Richard Whelan, notes that nurse-robots could help elderly patients remember to take their pills, monitor life signs, and alert medical services in emergencies. "Machines like this would allow people to remain independent and function on their own much longer," he says.

But robots may turn out to be more than mere doers of chores. Psychologists and sociologists have been tracking what appears to be a growing epidemic of loneliness in western society. And they suspect that many of us may buy robots to be our chums.

"I think the first commercially viable item may

turn out to be a robot pet, eventually even a robot lover. Don't forget the orgasmatron in Woody Allen's movie *Sleeper,*" says Stuart Lipoff. "But the first models would be fairly simple, with some artificial vision, some artificial voice, the ability to understand speech, some movement. It wouldn't have to do much more than move around, blink its lights, respond in a playful way, maybe wag a tail."

Soft and fuzzy, they could have built-in heating units, making them warm to the touch. Such robots might be therapeutic in nursing homes, where patients are denied pets.

Will people really choose machines to be their buddies? "Go back to *Star Wars,*" suggests Lipoff. "What were those two robot creatures really doing? They were not so much utilitarian as companions, friends."

You could do many things with your robot friend. Certainly you could play chess or checkers. Dr. Marvin Minsky, head of artificial intelligence research at MIT, years ago created a robot deft enough to catch a baseball. And, as with any friend, you could have long personal talks with your robot; people were delighted to discuss their problems with a computer "psychologist" created at MIT not long ago.

"If people already have trouble differentiating between their relationships with people and a machine, over the next decade or so, as we develop computers with ultrahigh-speed parallel processing, people may find conversations with a robot indistinguishable from talks with people," says Martin Ernst. "They may even find the machines preferable."

Here we are, hardly settled into our electronic cottages, and already the age of electronic pals is upon us.

As Nolan Bushnell told reporters when he announced his new protege, Androbot: "We're talking about a someone, not a something. A friend that would greet you after a long day at the office."

What next? Bushnell predicts synthetic travel. You enter a control module in Duluth and take command of a robot in Paris, peering through its eyes, listening through its ears, as you send it lurching down the Champs-Elysees.

"It's a wonderful business to be in," says Joseph Engelberger. "All things are possible.

BORN AGAIN ROBOTS

By Robert A. Freitas, Jr.

The next 20 years may witness the birth of a man-made life form that could lead us into space—and eliminate most human labor here on Earth. Much of the preparatory work toward it has already been done.

Picture one possible result:

From a rocket that left Earth several years before, an enormous egg drops to Saturn's ice-moon Enceladus and cracks open, releasing the robot inside. Stilting spider-like around the surface, the automaton sets immediately about reproducing itself, using only the materials at hand and feeble light energy from the distant sun.

Soon the robot and its descendents begin their real task: mining the Enceladean ice and building small light-sail tugs to carry the chunks toward the inner solar system. For a time, earthly astronomers see nothing unusual, but eventually a new ring begins to appear around Saturn, surrounding the old ones at about twice the distance. A cloud of robot vessels spirals outward from Enceladus until the sun's gravity balances Saturn's, then spills in a long stream toward Mars.

Their shipments of ice fall like sparkling meteors on the Martian surface, melting on impact and thawing the frozen ground. First rivers, then whole seas coalesce. The air grows thick and warm, and soon it rains on Mars for the first time in perhaps a billion years. Within a decade, human colonists arrive on their new world.

According to Dr. Robert A. Frosch, former NASA administrator, such missions are not only possible but necessary. In a talk before the Commonwealth Club of San Francisco, he told a startled audience that to support ourselves in space, we will need self-reproducing robots. They would, he declared, "provide easy access to the resources of the solar system for a relatively manageable investment."

The key to the scheme is a machine that can use solar energy and local materials to build a replica of itself, with little or no human guidance. Building generation after generation of offspring, the total number of machines would grow exponentially, the way biological populations expand. So would their output of manufactured products.

NASA is taking this concept very seriously. In 1980, it held a ten-week summer study session at the University of Santa Clara. My group, called the Replicating Systems Concepts Team, studied the idea of setting up a self-reproducing factory on the moon that would eat raw lunar soil and manufacture anything we need.

Our basic plan would put a 100-ton seed full of machinery on the moon. The first robots would emerge to fuse the lunar topsoil into a circular factory site of cast basalt 100 yards across. Then they would install the factory itself and erect a canopy of solar cells to power the system.

The factory has three major sections: One extracts purified elements from the soil, another forms them into machine parts, tools, and electronic components; and the third assembles the parts into useful products. In a year, a 100-ton seed could extract enough material to duplicate itself. If allowed to grow undisturbed for 18 years, the factory output would total more than 4 billion tons per year—roughly the current industrial output of the entire world.

A growing, self-replicating factory could be programmed to mass-produce robot miners and spacecraft—almost anything we need. "It could build a few thousand meter-long robot rovers equipped with cameras, core samplers, and other survey instruments," suggests Georg von Tiesenhausen, assistant director of the Advanced Systems Office at the Marshall Space Flight Center and a member of the replicating-systems team. "They could cover the moon like ants, mapping it in just a few years. By conventional methods, it might take a century or more."

How soon could such a system be in operation? Von Tiesenhausen says that within 20 years after the project is begun, the U.S. could produce the first robot able to duplicate itself from raw materials. Former NASA administrator Frosch is even more optimistic. "We are very close to understanding

229

how to build such machines," he says. "I believe that the technology is already available and that the necessary development could be accomplished in a decade or so."

Long before NASA became interested in self-replicating machine systems, the basic theory had already been worked out in some detail; much of it had been around for more than 30 years.

It began in 1948 with the late John von Neumann, a brilliant Hungarian mathematician famed for his early work on electronic computing. In a series of lectures at Princeton, he discussed how automata might reproduce themselves.

According to von Neumann, a self-replicating machine must have at least four distinct components: the builder, the copier, the controller, and the blueprints. Reproduction starts when the controller commands the builder to construct exact replicas of all mechanical systems, according to instructions in the blueprints, a sophisticated computer code. The robot would pick the right machine parts from its stockroom and assemble them in order. Then the controller would command the copier to duplicate the blueprints, insert the copy into the replica, and turn the new robot on. Voila—two machines!

Other scientists have also given serious thought to self-replicating automata. Physicist Freeman Dyson, of Princeton's Institute for Advanced Studies, suggests that a small robot adapted to earthly deserts might duplicate itself from the silicon and aluminum in the rocks around it. Powered by sunlight, it would manufacture electricity and high-tension lines. Its progeny could eventually generate ten times the current electric output of the United States.

Though the robot's potential for destroying the natural environment would be enormous, Dyson believes it would eventually be licensed for use in the deserts of the western U.S.—probably after bitter debate in Congress. In the end, the robots would probably have to carry within themselves a memory of the original landscape and restore its appearance whenever the site was abandoned.

"After its success here," he speculates, "the company that built it might market an industrial development kit for the Third World. For a small down payment, a nation could buy an egg machine that would mature within a few years into a complete system of basic industries, along with the associated transportation and communications networks." A spinoff, he suggests, might be the urban renewal kit, with self-replicating robots programmed to build new neighborhoods from the debris of the old ones.

Former A-bomb designer Theodore Taylor, now a consulting physicist at Princeton and head of the International Research and Technology Corporation, likes to call such devices "Santa Claus machines" because they could stuff our stockings with almost anything we ask for. Taylor himself envisions a space-based version that would turn out almost any product on radio command from Earth.

Scoopsful of moon rock or asteroid material would be broken down into a beam of ions and passed through powerful electromagnets. Since the electrified atoms swerve in a magnetic field and light atoms veer more than heavy ones, the material could be sorted atom by atom into its constituent elements. To make, say, a washing machine, the Santa Claus device would select the necessary materials, vaporize them, and spray them into mold.

Because the machine does all the work with little human input, the cost of goods could fall nearly to zero. If the workload grows too large, the automata simply reproduce themselves and go on working. The effect, Taylor says, is an "infinite degree of automation."

Computer scientists have received such schemes enthusiastically. Dr. Ewald Heer, a robotics specialist at NASA's Jet Propulsion Laboratory, in Pasadena, calls self-replicating robots one of the most fascinating ideas for the future of space. "This offers a way to create a self-supporting economy by robot labor," he observes. "Immigrants from Earth could set out with a minimum of risk, knowing that the means of their survival had already been provided."

Dr. Michael Arbib, of the department of computer and information sciences at the University of Massachusetts at Amherst, suggests that they might also be used for interstellar communication. "A self-reproducing machine might well carry out its own synthesis from the interstellar gas," he offers. These machines, Arbib says, could then reproduce in space, creating an expanding sphere of explorers moving outward in space.

With this scheme in mind, Dr. Frank J. Tipler, of Tulane University, has even argued that intelligent aliens cannot exist: If they did, they would have had to build such machines to explore and use the galaxy, and we would see glaring evidence of this all around us.

Some machines have already managed to reproduce in primitive ways. Self-replicating computer programs have been written in nearly a dozen different languages, and small machines that can copy themselves from simpler parts have proved remarkably easy to build.

One basic model was developed years ago by British geneticist L.S. Penrose at University College, London. It is an ingenious set of interlocking blocks with clever arrangements of springs, levers, hooks, and ratchets. A two-, three-, or even more-block assembly can replicate when placed in a box with other loose blocks and shaken gently. One end of the completed assembly hooks onto the loose blocks in the right sequence, building up a duplicate chain and then releasing it when the final block is connected.

Homer Jacobson, a physicist at Brooklyn College, in New York, built another such device using an HO train set. Two kinds of self-propelled modified boxcars called "heads" and "tails" are circulated randomly around a loop of track with several sidings. If a head and tail are first assembled on a siding, the pair can reproduce itself.

First the head tells the tail to wait for a new head car to come by and shunt it onto an adjacent siding; the tail signals the head when that has been accomplished. Next the original head tells its tail to shunt a new tail onto the siding and then turn itself off. The new tail latches onto the new head, turning the head

233

on and making another active couple. This second couple can reproduce using the next siding, and so on until all sidings or components have been used.

Such experiments sound much too simple to justify calling them reproduction—nothing like the mysterious processes that form a new human life. You might even object that von Neumann's whole concept is just a general-purpose assembly robot— whose output happens to be copies of itself. But "after all," observes Dr. W. Ross Ashby, a biophysicist at the Burden Neurological Institute, "living things that reproduce do not start out as a gaseous mixture of raw elements." Even human beings require a specialized environment supplied with air, water, and nutrients in order to procreate. Von Neumann's robots are just a little less independent.

In fact, some scientists already feel that computers are more than mere machines. Dr. John G. Kemeny, president of Dartmouth and one of the inventors of the computer language BASIC, believes that computers should be considered a new species of life. "Once there are robots that reproduce themselves," he declares, "it would be easy to program them so that each offspring differs slightly from its parents. It would probably be a good idea to let each robot figure out some improvement in its offspring so that an evolutionary process can take place."

But compact, self-reproducing robots still lie somewhere over the technological horizon. According to Dr. Marvin Minsky, head of artificial intelligence research at MIT, an automaton today would have to be the size of a factory to reproduce itself from raw materials rather than from prefabricated parts.

Fujitsu FANUC, Ltd., a manufacturer of numerically controlled machine tools, took a giant step toward that goal with a $40-million robot factory it opened in January 1981. Robots there are built by other robots, with only 100 humans to supervise and help in the manufacturing. The first "unmanned factory" in the machinery industry, the plant produced 100 robots and other electronic equipment in its first year. Once such a plant can make all of its own components, it can be programmed to make more of itself—to reproduce. (See "Robots of Japan," *Omni* January, 1982).

Since we cannot foresee all the problems these robots will have to face on a distant planet, we must supply them with goals and with the problem-solving ability to carry out their assignments in our absence. It seems at least possible that machines this complex will begin to evolve some of the social behavior common to animal populations. This brings very close the prospect of sharing our planet with a form of near-life whose evolution we cannot predict.

At the simplest level, what would happen if one machine begins to neglect its production chores in order to reproduce? Its offspring—possessing the same trait—might soon dominate the machine population. Would some form of "kin-preferring" behavior arise? Might the robots even develop a form of "reciprocal altruism" in which the machines behave in seemingly unselfish fashion toward others that are not "kin" in order to create a more stable "society?"

"If our machines attain this behavioral sophistication," notes Dr. Richard Laing, of the department of computer and communication sciences at the University of Michigan, "it may be time to ask whether they have become so like us that we have no further right to command them for our own purposes, and so should quietly emancipate them."

And one wonders: Could such self-reproducing robots someday become our enemies? The usual answer is that we can just pull their plugs to regain control over them. But is that really so? We are already so dependent on simpler computers that to shut them down would cause economic chaos. Of the Santa Claus machines, theologian Ralph Wendell Burhoe, of the Meadville/Lombard Theology School in Chicago, asks, "Will we become the contented cows or household pets of the new computer kingdom of life?"

And what if the machines learned to defend themselves? The Replicating Systems Concepts Team at the Santa Clara study session concluded that to escape human control, any machine must have at least four basic abilities: It must create new ideas to explain conflicting data, inspect itself completely, write its own programs, and change its own structure at will. A machine that lacked even one of these abilities would almost surely be unable to anticipate or prevent its own disconnection. It seems unlikely that machines will soon acquire these powers; and even if they do, it will be because human beings decided to supply them.

Nonetheless, a few people view the future gravely. Dr. James Paul Wesley, associate professor of physics at the University of Missouri, points out that the advent of machines has been amazingly abrupt compared to the billions of years it took

236

carbon-based life to evolve here on Earth. Yet, he believes, the same laws of reproduction apply to machine as to biological evolution.

"Machines," Wesley cautions, "have also evolved toward an increased biomass, increased ecological efficiency, maximal reproduction rate, proliferation of species, motility, and a longer life span. Machines, being a form of life, are in competition with carbon-based life." The result, he fears, is that silicon life "will make carbon-based life extinct."

Yet there is another possibility. What we are approaching, says NASA computer scientist Roger A. Cliff, is "cybersymbiosis"; eventually, humans could come to live inside a larger cybernetic organism. As man and machine evolve, our relationship with it will cease to be voluntary and become necessary. Cliff views this development with eagerness. Our descendents could live inside large, self-replicating, mobile space habitats which act as extraterrestrial refuges and guarantee that humanity is never wiped out by some earthly catastrophe.

"Flesh and blood are ill adapted to space," he comments, "but silicon and metal are ideal. Just as our own DNA resides within a protective membrane and mitochondria are locked within cells, so might humanity live as cybersymbiotic organelles of the space-colony organism. I see these traveling throughout the cosmos, searching for nutrients—asteroids, gas clouds, and so on—growing, evolving, and reproducing. And in their offspring will be us."

JAPANESE ROBOTS

By Bruce McColm

The silvery, six-foot-tall robot resembles a benign metallic, Frankenstein monster, with television-camera eyes, an artificial ear built into the stomach, and a deep synthetic voice. It's made of high-strength aluminum alloy, weighs about 130 kilograms, and comes apart easily: All last summer its long legs were stored in a laboratory filled with dis-embodied mechanical hands, arms, and computer components. But the robot's maker, like many modern Japanese, treats it as part of the family. Witness the home movies.

The films shot by Waseda University's Ichiro Kato, proud developer of the Waseda Robot, or Wabot, are like any rough chronicle of the early years of a new baby. There are the first two uncertain steps, requiring a full 110 seconds to complete. There is the later, smoother gait, covering the same ground in nine seconds. The soundtrack (this is a modern Japanese home movie) preserves some of the early words. "What is your order?" the Wabot asks an operator, who tells it to move a step to the right. And the machine does more than merely obey. "Now I start," it boasts.

In one scene the Wabot moves its upper torso, and Kato, the wispy and paternal chairman of the graduate school of science and engineering at Waseda, could not be more pleased if he were watching one of his students reinvent the transistor. "There you have a one-and-a-half-year-old toddler," he says. And he dreams a father's dream about what his precocious toddler might become as it grows up.

"When a child becomes seven years old, he has an ability equal to the adult. My goal is to build a robot that will have a seven-year-old's ability. Even though it's not possible, I'm trying to do it."

Kato is one of a number of Japanese scientists working on the generations of robots to come of age in the twenty-first century. They will be quite different, he predicts, from the rivet-punching drones of today. "The industrial robot today is the mainstay of all robots," Kato said over tea. "They are simply laborers or function oriented. But in the next twenty years we will witness the emergence of the robot in service areas such as medicine. The robot will take over in everyday situations and will closely resemble human beings with the integration of artificial intelligence, voice, tactile recognition, and bodily functions. Through this mechanization of the human being, we will know more about ourselves. And to know yourself is the long pursuit."

Ever since the Czech novelist Karel Capek coined the word *robot,* man has dreamed like Professor

Kato of creating an intelligent machine to liberate him from the drudgery of work. No other country has pursued this vision quite so vigorously as Japan. Since it first imported an American-made industrial robot in 1967, Japan has emerged as the robot center of the world, the model of future societies.

Today there are more than 77,000 industrial robots in Japan, nearly 70 percent of the world's robot population, most of them tucked away and turning out products in factories. But thousands of new robots never go to work. There's a Japanese cat robot for sale, for instance, that uses artificial vision to avoid obstacles. Masked and menacing robots (like the one on the previous page) compete with humans in the sport of kendo, thrusting and parrying with pliable bamboo swords. Students at a Tokyo medical college give mouth-to-mouth resuscitation and injections to a robot programmed to suffer cardiac arrest on demand. Its soft skin sweats. Its pupils dilate, and its pulse trips appropriately. A monitor (in the photo above) provides a readout of its health, which can be instantly reset to normal.

The craze for machines like these is only just beginning. In 1980 Japan's businesses invested nearly $3 billion in robotization. Japanese experts predict that in a mere three more years total investment in robots will be at least 12 times as much. This is the kind of mammoth expenditure that might be expected to make some people queasy, raising a specter once presented by Isaac Asimov that humans are the first creatures capable of building their own replacements. But Japanese futurists are quite smug about the prospects of turning robots loose on the streets.

"In Japan we have a tradition of having things similar to slaves in the village, such as a horse and a cow," says Sakyo Komatsu, future planner and science-fiction writer. "These were considered a part of the family. When machines such as the automobile and electric trains were introduced into Japan, they were considered living creatures. Even in the villages today, especially during the New Year celebration, you can see cars being decorated with paper ornaments. So, for the Japanese, to witness this emerging of robots is like watching small children who are there to help their parents."

The motif was repeated during *Omni*'s visit as often as the musical theme from *Star Wars,* piped over Muzak: Robots are good for the family. Tokyo TV station NHK broadcast one typical robot success story in a documentary about a mom-and-pop manufacturing firm. Facing bankruptcy because of the cost of labor, the owners rented two robots and hired their son back from a job in Tokyo to automate their parts-making plant. In an offbeat switch on Alvin Toffler's electronic cottage-industry idea, the family was reunited, the business turned a profit, and the mother liked the robots because she didn't have to serve them tea all the time.

Bigger Japanese companies are in many ways simply extended families, with plenty of room for exotic mascot machines. "Instead of the great feudal families," one government employee says, "we now have the large corporations like Mitsubishi that protect the country and their own workers." In times of recession or runaway inflation, these companies improve industrial efficiency

242

instead of laying off workers: They buy more robots. Like Samurai warriors, who value honor over life, these machines work tirelessly, make few demands, and go unflinchingly to pieces when newer and more efficient generations of robots come along to replace them.

They also make a lot of money for their masters. Productivity in the Japanese auto industry quintupled—from a daily rate of 5 or 6 cars per worker to 30 or 40—with the introduction of industrial robots. The Nissan Motor Company, in Zama, about 35 kilometers south of Tokyo, turns out 1,300 cars a day, with 150 robots performing the work of 300 men.

Attracted by figures like these, dozens of companies outside of Japan's robust automotive industry have put robots to work. Machines made by Fuji Electric Company now sort out defective drugs, grade fruit, and crate eggs. Hitachi, besides making its own robots, uses them to assemble vacuum cleaners and other appliances, and Mitsubishi and Kawasaki are developing robotic divers to inspect deep-sea oil lines. And last January Fujitsu Fanuc opened a $38 million plant beneath the slopes of Fujiyama, where robots produce *other robots* in a factory coming close to full automation and the ability to operate nonstop.

Western visitors are often surprised at the warmth between Japanese people and the proliferating machines. Workers tag robots with the nicknames of movie stars and rock singers and remain fascinated with devices that frequently outperform them. By contrast, many American bluecollar workers are fearful of losing their jobs to hydraulic muscle and cold circuitry. The fears have some foundation: By some estimates, the number of blue-collar workers in the U.S. auto industry will decrease as much as 25 percent by the year 2000 because of robots. Japanese industrial experts are well aware of the potential problems of transferring tasks to robots too fast.

"An all-robotized environment is not necessarily healthy," says Yukio Hasegawa, a professor at the System Science Institute of Waseda University. "While workers, particularly in the auto industry, have good relationships with the robots, if you decrease the number of the work force too rapidly, the workers may get demoralized. If you decrease the number of workers from sixty to twenty, for example, the remaining work force might feel surrounded by robots, which are then seen as their competitors."

But to some extent in Japan the problem solves itself, because there aren't enough laborers: The government estimates that Japan currently needs some 840,000 more skilled workers, mostly for smaller enterprises. The shortage is likely to become more severe, according to Kanji Yonemoto, executive director of the Japan Industrial Association. By 1985 about 60 percent of the Japanese work force will be involved in service or information-oriented industries, not in production work, Yonemoto says. Most young people won't enter the pool of skilled labor because 60 percent of them today are attending universities, headed for white-collar offices. The situation has made it somewhat easier for robots to enter larger factories, and smaller enterprises require them desperately.

It is against this backdrop of familial acceptance and economic need that researchers like Kato are planning the intelligent robots of the twenty-first century. And in their work they're challenging the stereotype of the robot as a clunky, bland-voiced android.

In one project at the Tokyo Institute of Technology, Dr. Shigeo Hirose has built a series of intelligent robots that walk on four legs and propel themselves like snakes. Originally the snakelike robots were built simply as interesting experiments. But today they are being manufactured as industrial machines, and they may serve a wide variety of functions, from inspecting nuclear-power plants to moving patients around in a hospital.

Abandoning the human metaphor often used in robotics research, Dr. Hirose spent five to six years studying the movements of snakes. "We thought of making robots by taking living organisms as an example," Hirose recalls. "Human beings are too complicated. So we looked at animals; we thought the snake would allow the robot to have a wider function than it does now."

The results are the "activated cord mechanism," or ACM, a meter-long articulated pipe that may be twisted at joints to form any shape. Computer-operated cords inside the pipe act like tendons. Someday these sinews may be used for aiming an endoscopic camera at the end of a tiny robot injected into the lower intestines. Another snakelike robot is the "soft gripper," which can grasp an object of any shape or hardness. It may be called into service soon by the Tokyo Fire Department for rescuing people from burning buildings or places filled with poisonous gas.

Hirose moved on to a boxlike stucture supported by four spindly legs. "Here I was inspired by the movement of the spider," he says. The robot creeps, guided by a laser sensory system that fires 100 light pulses per second to provide a picture of surfaces up to a meter away. The next generation of the spider will carry 20 kilograms, and Hirose predicts that ultimately it might replace the baby carriage.

In Ibaraki Prefecture, some 60 kilometers northeast of Tokyo, is Tsukuba Science City, a future metropolis of some 200,000 people, mostly scientists, built in 1979 by order of the Japanese government. There, amid the farmlands and the creeping suburbs of tract houses with pagoda roofs, Dr. Eiji Nakano, the director of robot engineering, is working on a host of projects to meld artificial intelligence with the brute strength of the robot.

Some foresee a time when teams of mechanical spiders will cart fallen trees, harvested by robot lumberjacks, to automated mills. Humans will have little to do with the logging operations but plan and think.

Perhaps the most novel development in Dr. Nakano's Mechanical Engineering Lab (MEL) is the Japanese version of *My Mother the Car,* an automobile steered by microprocessor. The robot auto uses television cameras for eyes, and it brakes and accelerates on cue to miss oncoming traffic. "It won't be very practical for at least twenty years," Nakano says, "because it will demand such a social investment. Perhaps it will be useful for long-distance driving. But in Japan, as elsewhere, driving is after all a very personal affair."

Other MEL projects are prototypes for aids to disabled people. One is a small vehicle on wheels, with a microcomputer brain and ultrasonic eye, the main components of what might become a wheelchair to carry a handicapped person through narrow corridors. Another MEL project is a robotic guide dog for the blind, a small, scooting machine that warns of obstacles detected by ultrasonic sensors. Initially the robot is "taught" to use local landmarks to guide its master, but in the future the blind will be able to program the robot automatically with directions for local errands.

Touring the grounds of Tsukuba is a lesson in specialization. In the electrotechnical laboratory, across a highway from MEL, Dr. Seiji Wakamatsu and his colleagues have developed one robot capable of sawing wood and building a box, another possessing the dexterity of the human hand, and still others capable of sensing moving objects.

While Dr. Wakamatsu believes that robots will continue to be developed according to their projected industrial use, Kato believes that all-purpose robots imitating humans in intelligence and bodily functions are more necessary. "There is a constant theme in science fiction that robots will destroy human beings," he muses. "They do so only at the point when they acquire emotion. If I consider emotion part of intelligence, then I don't think we can build a robot with equal ability. But for now, the more versatile a robot becomes, the more it resembles a human in shape.

"When we come to a service robot, the shape will have to be similar to that of a human being. In the case of the factory, you can lay out the plant in such a way as to make it suitable to the working conditions of the robot. But when robots come into our houses, if we change our house for a robot, then it really should be the other way around."

A colleague of Professor Kato's, who shares many of his ideas about the future human robot, is Shunichi Mizuno, a robot artist who works in a small studio and workshop in the suburbs of Tokyo. His creations are Disney-like characters, exhibited all over Japan, most recently at Portopia, the scientific exposition held on a man-made island off Kobe, in southern Japan. For the Tsukuba Expo in 1985, he is building a Robot Theater, presenting intelligent robots equipped with pattern-recognition devices, voice, and sensors to form dancing chorus lines like the Rockettes. An earlier creation elicited a mild sensation in Japan: Mizuno constructed a robotic Marilyn Monroe, which played a guitar and danced on national television.

"In robotics, art and technology converge," he asserts while pulling the vinyl face off an old man built to promote solar energy. "Ultimately we can have Marilyn Monroe sitting at a reception desk and answering our telephones. Robot actors may replace Faye Dunaway. Basically, if a robot is an entertainer, it's human, with all the appropriate facial expressions. Once we begin to use the human form and place a computer in it, we are finally raising questions about the definition of life. If I thought about that very long, I'd be afraid to make robots."

But Mizuno will not stop making them, not even if robots become indistinguishable from people.

Already robots are on the verge of taking over manufacturing, from the processing of raw materials to the final storage of the product in an automated warehouse. Soon coal miners will be replaced by snakelike robots that burrow into the earth, controlled from a command center aboveground. Diagnostic robots, propelled by magnets, will inspect nuclear-power plants, and their cousins will paint the sides of ocean liners. Robotic migrant workers will be sent to the ocean's bottom and into space as Earth's natural resources become depleted.

One afternoon in Tokyo Masaya Nakamura, president of Namco, Ltd., a maker of computer games and robots for promotional campaigns, gathered his staff around him. It was a tradition at the end of the day. The staff listened, and Nakamura mused. "What Toffler said about the Third Wave is beginning to be true in Japan," he told them. "The development of technology is meant to bring happiness to human beings. Machines are a part of us, like a partner. When you put a coin into a pinball machine, you know how it feels. It is not just a machine, but just like us."

PART SIX:
DIGITS AND DOLLARS

MR. CHIPS
GOES TO
WALL STREET

By Anthony Liversidge

What was easily one of the most entertaining and puzzling news events of last year was listening to David Stockman, the *wunderkind* director of the Office of Management and Budget (OMB), trying to explain away 11 months of damaging admissions, which were eventually published as an article in *Atlantic* magazine. Barely three months into the Reagan presidency, Stockman privately had lost faith in the ability of the President's proposed tax cuts to stimulate economic growth and avoid huge budget deficits, even as he publicly rejected such fears and pushed hard to get Reagan's program accepted by Congress.

Here was the leading architect of U.S. economic policy, with all the resources of the federal government at his disposal, completely unable to make any kind of forecast as to what his economic policy would lead to. None, anyhow, that he could stick by for more than a couple of months. Here was the director of one of the power centers of government, with the most thorough, most up-to-date information on the state of the economy, admitting, after a detailed analysis of all available data:

"None of us really understand what's going on with all these numbers!"

At the heart of this confusion was a relatively new approach to economics, called econometrics. Very simply, what David Stockman had done was feed the OMB computer a program known as an "economic model," a set of instructions that would make it mimic the behavior of the economy in the real world and play out different economic scenarios according to what policies he wanted to try out.

In the past decade this computerized approach to economic prognostication has become more and more common both in business and in government. The trouble is, as the Stockman drama suggests, econometric forecasts are far from infallible. In fact, how much they add to the accuracy of forecasting economic swings is still hotly disputed. Just the same, hundreds of millions of dollars and some of the best brains in the economic field have been invested in this new approach; for many, forecasts made with the help of computer chips are crucial to our economic future.

Econometric models came into common use 40 years ago at the University of Chicago. There a young economist, Lawrence R. Klein, had built a model, or mathematical blueprint, of the U.S. economy. With it he predicted correctly that the

United States would not sink back into a depression after World War II—something most economists feared at the time. In the Fifties a slightly more developed Klein model again beat the consensus by projecting, not a severe, but a mild recession after the Korean War. And by 1963 Klein, then a professor at the University of Pennsylvania's Wharton School of Finance, began offering his annual and quarterly forecasts to business and government clients for charting economic strategies.

Klein's early models were simple, with few equations, partly because computations then had to be done by hand. Once hitched to the computer, however, the modeling bandwagon really began rolling. The combination of its hocus-pocus of economic theory, its dazzling mathematical expressions (incomprehensible to the layman, not to mention many corporate chief executives), its electronic wizardry, and an aura of computerized precision became altogether unbeatable.

By the end of the Sixties, corporations, universities, and government agencies were using models. Wharton's list of clients had by then increased from 5 to 70. The moneymaking potential in consulting was very quickly evident.

The leading rivals to Wharton were born, first, in 1969, when a Harvard economist named Otto Eckstein founded Data Resources, Inc., or DRI, and again, in 1970, when Michael Evans, a colleague of Klein's, left to head Chase Econometrics Associates, funded by Chase Manhattan Bank. These Big Three—Wharton, DRI, and Chase—dominate a forecasting industry whose clients routinely pay from $20,000 to $100,000 or more annually for access to the firms' model forecasts and other services. Today model forecasting is a $100 million industry.

Of the Big Three, the biggest is DRI, which serves more than half the market. DRI provides its hundreds of clients with short-term and long-term models of the American and of most other developed nations' economies as well, and its computer terminals can tap in to an economic and financial data base that is enormous. Three years ago the company was bought by McGraw-Hill for $130 million, netting Eckstein some $13 million.

The Wharton model was swallowed by another hungry conglomerate, Ziff-Davis Publishing, in 1980. Klein, now sixty-one, is said not to have profited much personally from the deal and in fact was not interested in what he calls "big bucks." But public recognition came to him that same year when he won the Nobel Memorial Prize in economics for his pioneering work—a sign that econometrics had truly arrived as a science.

So why, then, did a colleague of Klein's wisecrack in 1980: "Another year and they'd have to give Larry the prize for mythology"? Because in the last few years the econometric futurists have not been all that accurate in their predictions.

For example, when an Arab oil embargo led to a recession here in 1975, nearly everyone kept predicting economic expansion until the decline was virtually upon the United States. And again in 1977 nearly all models predicted a recession for 1978 and 1979—which didn't happen. In the last two years there have been quarterly fluctuations in the gross national product, or GNP, which model forecasters conspicuously failed to foresee. And the worst quarter forecasts were for the winters of 1980 and 1981, when all the models projected a decline in U.S. economic output for the first three months of the year. What actually happened was that the GNP had a mighty surge; it didn't take a dive.

Consequently, there has been some disenchantment with economic forecasters. In professional and business circles, when models are mentioned, the skepticism is pervasive. At the corporate headquarters of American Express, in Manhattan, for example, the firm's chief economist, John J. Casson, says of econometric models: "We are very dubious about how good any of them really are."

And on Wall Street, Barton Biggs, a senior investment manager at the prestigious investment banking firm of Morgan Stanley, states flatly, "They haven't worked very well. That's the record."

Academic critics can be even more acerbic. The influential monetarist Karl Brunner describes large models as "numerology on the same level as astrology." He says they use unscientific hypotheses that can't be checked against any results. Large models, he adds, have had "a miserable record."

However, the econometric models' commercial appeal remains strong. Despite their skeptical remarks, both American Express and Morgan Stanley still subscribe to the chief model forecasts. They even have their own in-house versions as well. And politicians are not immune to their appeal, either. President Reagan's new economic policies were formulated by using at least two models so as to justify their expected outcome, namely, a recovering economy.

Part of the models' popularity comes from what one economist calls their "scientific mumbo-jumbo." A large model basically is a long list of complex-looking equations. For example, DRI's short-term macromodel, mirroring the month-by-month workings of the whole U.S. economy, uses 1,200 of them. A computer printout of DRI's model equations is as big as a hefty book, measuring 8 inches wide, 11 inches long, and about 2 inches thick. Just *one* interminable equation, representing the value of American oil imports, takes up 82 lines. The first line reads: "1 > MEND1067 = (EXP (< 42.1073 > + < 0.980795 >*LOG (DTFUEL-SALLB*PET&NG%ENERGY . . ." And the technical documentation explaining all of this runs an additional 400 pages.

The economic theory behind the model seems equally impenetrable to the uninitiated. "Typically the investment equations are based on the Jorgensonian Neoclassical Investment Function," says Chris Gutry, director of the macromodel used by Merrill Lynch Economics. "The consumption functions are some sort of permanent income hypothesis, sometimes modified by real cash balances. Uh, are you an economist? I don't want to bore you."

There is actually less here than meets the eye. The model language is a simple mnemonic, a retrieval code designed to help make the model accessible. DTFUELSALLB, for example, translates: "Demand for All Fuels, Total, All Sectors." Some equations are very short, and those that are long are not mathematically complex, not even for a layman.

"The mathematics are trivial, really," says Chris Caton, director of DRI's long-term model, "high-school level." The most complex terms, he says, are logarithms, and the operations are merely adding, subtracting, multiplying, and dividing.

The economic theory is not all that esoteric, either. "Conventional macroeconomics," Eckstein declares. "There is no central theoretical construct from which we derive everything else. The equations are each designed to be as good as we know how to make them."

"To calculate, say, the consumption of gasoline," Klein explains, "we let it be determined by a bunch of things related to automobiles and gasoline." In other words, there is more practical common sense involved in this than abstract theory.

In operation, a model can be a simple toy to play with. Any DRI subscriber can plug the nearest telephone receiver into a portable-typewriter terminal and be connected with DRI's Burroughs giant computer complex in Lexington, Massachusetts. He is, in effect, instantly, transformed into President Reagan or, better, Paul Volcker, chairman of the Federal Reserve Board—able to ordain a monetary policy for the United States and thereby help or harm the economic fortunes of millions of fictional consumers.

Chris Probyn, DRI's director of model development, provides a demonstration. Sitting at a conference table in the company's New York office, he inserts a telephone receiver into a terminal's rubber cups and types in a request for the macromodel's current gross national product forecast: "P 81 TO 82 %CH GNP72, PGNP, CPIU, RU."

Snap, whrrr, snap, whrrr, snap, whrrr. A table appears. It shows the projected GNP, consumer prices, and unemployment figures for 1981 and 1982. Mild recession till autumn, it appears. Then growth resumes. Hmmm. The President's dream. The "rosy scenario."

"Ask it what the budget deficit will be," he is requested. *Whrrr, snap, whrrr.* Uh-oh. The deficit fluctuates around $70 billion through 1982. Not so good.

"What about easing monetary policy?" *Whrrr, snap, whrrr, snap.* Oops! Inflation rises, but unemployment drops, real growth rises, fixed investment rises, and housing starts jump. Hey, does Paul Volcker know this? Apparently not. As this is written, money remains excruciatingly tight, and the economy has plunged into a recession with little relief in sight.

This kind of immediate interaction with the model, exploring alternative scenarios, solves what has for centuries been a serious problem for economists: having a laboratory situation in which they could test their economic theories. The real economic world doesn't stand still long enough to be dissected. Even if it did, it is far too large and complex to disentangle cause and effect with any certainty. Given all this, a computer model is a dream come true, a working microcosm of the economic universe to fool around with.

But a model alone is not enough to make accurate forecasts. It has to be steered. Human judgment is involved. "You don't just push a button and let the model churn it out," Caton explains. In addition to models, his firm employs 25 full-time economists in its National Forecasting Group, and the forecasts ultimately are made, not by the computer, but by the staff.

"Models run blind don't work," Eckstein asserts. "Judgment comes in interpreting why they didn't work perfectly. What was it: data errors, labor strikes, bad weather, changes in the structure of the economy? What is the government doing? Most successful forecasters have Washington experience."

Do you also need some kind of intuitive ability? "Sure," Eckstein answers. "What you have to be is a 'forest' economist, not a 'tree' economist. You must have a model in your head of how the whole economy hangs together. If you don't, even with the most wonderful model in the world, you'll probably turn out nothing but nonsense."

So much for the computer's aura of omniscience. But if a model must be helped along, what exactly *does* it do? It imposes discipline and consistency on the thinking of the staff, for one thing, and it acts as an extension of the economists' judgment. If a model cranks out a forecast that doesn't seem right to its masters, they simply override it and rely on human expertise.

Recently some new ideas have split the world of economic modeling. One is a new economic theory. Michael Evans, who now runs his own Washington, D.C., consulting firm, Evans Econometrics, uses a model that incorporates supply-side economics, the fashionable new theory that buttresses President Reagan's fond hope that massive tax cuts can boost national growth without causing inflation. Evans's model does *not* include the "rational expectations" theory, another idea employed to justify Reaganomics. This theory held that high interest rates would fall early in the new era, because Wall Street would "expect" the new policies to succeed. Since interest rates have stayed obstinately stratospheric, "those guys," Evans says, "are dead in the water."

Both theories were designed into Californian Claremont Institute's model, which Stockman and other Reagan advisers drew on in charting the President's economic course. Or so John Rutledge, Claremont's president, avers. No outsider can check this out. Rutlege has fended off any independent evaluation of his model by keeping its design secret.

Rutledge also believes in another trend in econometrics: small computer models. He insists one big advantage of the small model is its agility. "Ours is simpler and more flexible," he says. "The large models are juggernauts that simply can't react fast enough to changing conditions. They are inherently doomed to fail."

Evans now uses a model that is much smaller—having only 200 equations—than the one he designed for Chase. "I don't think the human mind can absorb more than that," he suggests.

Casson, of American Express, agrees: "The large models have gotten so huge that no one person in any organization can wholly understand them." His own department has designed about 20 models for its own use. They are small. How small? "Tiny. One equation," he replies, "with maybe five variables."

One area where tiny models have proved especially successful is foreign exchange. A recent survey of 12 currency forecasters in the financial magazine *Euromoney* found that smaller models often produced forecasts better than the bigger, more complicated ones. The best were the single-equation models.

Ranked first was a New York company, Predex, which has consistently beaten all the big banks in obtaining accuracy. Its founder, Dr. Charles Ramond, boasts that he has never taken a course in economics. His Ph.D. is in psychology.

"I believe in simple models," Ramond says. "We use a single equation, which simply states that the value of one currency in terms of another depends on the supply and demand for both. Every quarter we refit the model with the very best new data, and every night we offer silent prayers to God that the structure of the system won't change."

The Predex forecast is usually—70 percent of the time—on the "right" side of the currency forward market. That is to say, it makes money for its roster of clients, a list that includes Gulf Oil, Xerox, a slew of banks, and even the savvy Wall Street commodity house and money trader Salomon Brothers.

But small is not always beautiful. Predex's first forecast wrongly predicted a 1975 rise in the value of the dollar (it fell dramatically), and Predex's Canadian-dollar predictions were poor until recently, when the company built a special model for Canadian economists.

Like elephants swatting flies, Eckstein and the other forecasters who use large models flatly reject the criticism by smaller rivals that their theories are insufficient or outdated. If the big modelers' forecasting record is sometimes embarrassing—and

Eckstein concedes, "I wish it were better"—there are three good reasons.

First, the fuel on which the models run is impure. The raw data fed to the models are flawed economic statistics, which the federal government is consistently revising. "It's a tremendous frustration," says Eckstein. "Every year, when they restate history, we have to reestimate the model at tremendous effort and cost—redo the whole thing all over again."

Second, there are the "exogenous variables," input data that can't be derived from past statistics. Political, and sometimes even economic, factors can only be guessed at: War may break out; Arab oil producers may raise prices; federal spending may change; the money supply may expand or contract.

"Two thirds of our error involves doping out what Volcker is *really* doing, what Sheikh Yamani is *really* up to," says Eckstein. "That part of the error is irreducible."

Third, there is the shifting structure of the economic world. To keep pace with big changes, models have to be continually reengineered. The sudden ascendancy of OPEC, or President Reagan's curtailment of the welfare rolls, to cite just two examples, makes it a whole new economic ball game.

Unquestionably the less stable the real economic world is, the harder it is for models to project its path accurately. Given the current increase in economic instability of recent years, Eckstein considers it a marvelous achievement that forecasts have not gotten dramatically worse. Ultimately, he sums up, "No one knows the future, only God. And He's not in the business."

As 1982 began, *Omni* asked four top econometric forecasting firms to predict growth and inflation in the spring and summer quarters of the year. The shape of this outcome will first be discerned when the Commerce Department makes a preliminary announcement of the results on the twentieth of the month following the end of each quarter. A month later it publishes revised figures.

Conveniently for the less accurate seers, this is not the end of the scoring process. Further revisions may follow a year or more later. Last year the GNP figures for the entire *decade* were readjusted, necessitating a full-scale retrofitting of the models to conform to this new historical pattern. There's always room, therefore, for an inaccurate prophet to claim that the model's predictions will be vindicated when the official figures are corrected.

Precise numbers are not the key targets of the model forecasters, however. Most important is the question whether they accurately gauge the turning points of the economic cycle: when boom begins to falter, or when recovery gathers steam.

All the firms forecast a decline in output during the first quarter of 1982. According to these figures, all are expecting a turning point in the spring, when they expect America's economic growth to resume.

Forecasting Firm	Economic Growth in United States (Annual rate in percentages)	
	April-June	July-September
Data Resources, Inc. (DRI)	+1.5	+5.5
Wharton Econometric Forecasting Associates	+2.3	+4.7
Chase Econometrics	+3.1	+4.7
Merrill Lynch Economics	+4.2	+6.0

	Inflation in United States Consumer Price Index (Annual rate in percentages)	
	April-June	July-September
Data Resources, Inc. (DRI)	+6.6	+6.9
Wharton Econometric	+8.1	+7.9

Forecasting Associates		
Chase Econometrics	+8.5	+9.2
Merrill Lynch Economics	+6.3	+6.4

Can we trust these figures? Klein, the originator of it all, insists that model forecasting on the whole has been successful and is getting better: "When I started forty years ago, I'd have said what we have achieved today was out of sight. We never expected to do this well. To project U.S. total production a year ahead, plus or minus one or two percent on rates of change, is extremely accurate, I think.

"In 1973 we forecast a world recession of about the right order of magnitude as soon as the oil embargo was announced, against the contention of President Nixon. In 1979 we got the right movement in real GNP each quarter. In 1980 we did forecast a lightweight recession, and we were rather good," Klein says gloatingly.

The dominant question is whether that performance, still imperfect, can be improved further. Klein is doubtful. For one thing, there is the problem of the revision of basic data by the Commerce Department. What's more, he notes, the official statistics leave out the underground economy—the unrecorded transactions of the weekend painter, say, who accepts cash or barter to avoid paying taxes. Estimates of its size range up to 10 percent of the economy.

Paul Samuelson, of Massachusetts Institute of Technology, who wrote an elementary textbook on economics that made him a millionaire and who won the Nobel Memorial Prize for economics in 1970, thinks that forecasters sooner or later will come up against a wall of uncertainty, a limit beyond which they will be unable to reduce error.

"They'll reach a barrier, along the lines of the physicists' Heisenberg Indeterminacy Principle," Samuelson predicts. "When you try to guess next year's inventory investment, for example, you have to realize that God Himself hasn't made up His mind yet."

Model forecasting may not have reached that point, but Samuelson thinks that even if it has, the current level of inaccuracy is not intolerable: "It's an allergy rather than cancer." He is also sure that models have contributed much to the reliability of forecasts: "The computer methods would have to be reinvented if they disappeared."

Experts in the business generally seem to be optimists on counts other than accuracy. "We may not be able to improve accuracy much," Klein says, "but we should be able to give more information, over a long time, with faster reporting and a wider range of variables."

One thing that should help is the steady expansion of computer capabilities. Not only does this mean that the quality of the data—the raw economic statistics—that models use should improve, but it also means that the models will be able to handle and correlate more raw information.

If so, econometric models may come closer to satisfying such critics as Wassily Leontief, the 1973

Nobel Memorial Prize-winner in economics.

The kind of real-time-systems approach he envisions is already largely in place in Norway, Leontief says. "Believe it or not, they just held what we believe will be the last census there. You know why? Because the data now flow in every day. A completely computerized system absorbs a daily flow of information on technology, demography, monetary operations, natural-resources deployment, and so on."

Such a setup naturally suits a planned, socialist economy better than a free-enterprise one. Eckstein reports that a Soviet delegation visiting DRI was visibly excited by the potential application of computer technology to economics.

A high-level Russian official asked Eckstein whether he could put 50,000 economic units on DRI's computer, all interacting in real time. The Russians apparently had divided their economy into 50,000 factories and other production units, and they dreamed of connecting them all to a central computer. Then, the planners felt, they could really run the economy.

"We had to tell them that the current technology allows only two hundred units on the computer simultaneously," Eckstein recalls, "and that fifty thousand was quite a few years away."

There are even suggestions that economic data could be improved by aerial photography. Spy cameras now used to monitor Soviet military movements could study the volume of U.S. transportation traffic or agricultural crops.

Even now the growth of electronic transactions, from supermarket checkout to check cashing, promises better data input for the American models. A cashless economy, where most transactions are electronically performed, might provide the same kind of instantaneous picture of the economy that Soviet officials hope for.

With this potential, new complications arise. The prospect for DRI, as the nucleus of this computerized economy, would be fantastic, Eckstein believes. DRI's forecasts could be radically improved if Eckstein were prepared to tap directly into his clients' own plans, say, for investment spending.

But there are serious antitrust problems. "Do you really want large companies to pool their information into a computer?" Eckstein asks. DRI leaves it to the government to decide, because you have to ask yourself whether you are tampering with the system. "Is this fair to small business?"

The current outlook is that econometric models will be important tools in forecasting but will not magically replace human judgment.

Recently one of Samuelson's colleagues found him reading a computer printout and asked him, "Paul, in how many years will a computer make you obsolete?"

"Since he asked," Samuelson says, "I told him. 'Not in a hundred years.' "

ALICE'S FACTORY

By Kathleen Stein

Ozumba went upstairs to his study and punched in the code for the gear cluster he had begun to design. He had a couple of hours before the four o'clock satellite passed over his part of Nigeria. The cathode-ray-tube (CRT) screen filled with lines of color. Ozumba rotated the shape on its X and Y axes, reworked the teeth angle on a sprocket, ran stress analyses and sectional and mass-properties calculations. He simulated its action on the man-powered aircraft the gears were designed for. Earlier that day from his materials data base he had determined that the gears would be composed of a new titanium alloy. He zoomed in on the hub, reconnecting some lock rings. At five to four he checked the satellite dish, and when the bird was in range, he beamed up the program for the gears.

Ozumba's design was received at Widget International's South Orange, New Jersey, plant. It came up on the terminals of the structural engineers who made some modifications and sent the design to the mainframe computer, which made up materials orders and sent them to supply; alerted the computerized batch-order processors, which determined the number of gears to be made; found out which shop was available for the job and put its robots on call. The design was routed down a series of increasingly simpler computers that led to the microprocessors controlling the machine tools, robots, and sensors. The titanium sheets were delivered in automatic carts to work stations, where robots handed them to a variety of cutters and forges. Then robots transferred the finished gears to assembly areas, where they were put together and finally given over to stations where robots fastened them to the aircraft's main chassis. Inspections by sensor robots were run at every stage.

Another far-out scenario? Yes and no. Far-out because the implications of a computerized, workerless manufacturing system shake the foundation of what is known about industrial society. Not far-out because it is happening now. Some call it the Second Industrial Revolution; some say it's the most important technological innovation since electricity.

It's usually called by its acronym, CADCAM, which stands for Computer-Aided Design and Computer-Aided Manufacture. There are other acronyms, too: CAE, for Computer-Aided Engineering; CIM, for Computer-Integrated Manufacturer and more. They all bespeak an industry wherein the computer's electronic pulse ripples through the innermost recesses of production. Such huge corporations as General Electric have extensive research projects for setting up factories of the future, and the Robotics Institute of the Society of Manufacturing Engineers asserts that by 1990, 50 percent of all manufacturing work will be handled by robots—more than ten times the current level.

Fortune magazine calls CADCAM the "best new technology for productivity ever to come on the market." Indeed this radical new equipment promises a lifeline to an American economy that Treasury Secretary Donald Regan calls "dead in the water," even as it offers the possibility of production improvements as great as 30 to 1. For better or worse, CADCAM is changing the nature of blue- and white-collar work forever. In tomorrow's factories all the workers will wear titanium-alloy collars.

The CAD part of the equation refers to the use of computer graphics (and accompanying data bases) to create two- and three-dimensional pictures from hypothetical realities—kinetic neo-blueprints for designing everything from Halloween masks to microchips; from Black and Decker hand tools to four-cylinder engines to shuttle skeletons (see illustrations on previous pages); from tractors and airplanes to petrochemical plants. CAD is used in pharmaceutical research to construct new molecular compounds; in simulated testing of parts (like the turbine components above); in architecture to design skyscrapers. In the aerospace and automotive industries CAD has been in place for more than a decade. One General Motors designer admitted, "We wouldn't know how to do it manually anymore."

The CAM side of the loop—the computerized control of production machines from numerically controlled (NC) milling devices running on punched tape programs to semiintelligent robots with visual and tactile sensors—is just beginning to penetrate industry. And its applications are equally far-reaching. Although factories have had NC machines for more than 25 years, they have been serviced by humans rather than by robots and computers. Increasingly, though, minicomputers are taking over from people—measuring parts, calculating tool paths, piloting the robots, directing finishing touches on a product. Computers can also preside over foreman-level jobs, sending out communiqués about batch-size changes, bills of material, shipping, and even the self-repair of shop equipment.

CADCAM, this automated "great chain of being" for industry, transforms the factory into a kind of living organism linked together by intricate feedback of electronic messages. In time the "memories" of this organism will accrete in vast "Expert Systems," digital repositories of shared wisdom about the plant and its manifold processes.

At the top of this chain is the engineer seated before his CRT screen. Looking over the shoulder of one of these masters of the electronic drafting board, you might see an image on the screen that looks something like a Mondrian painting—except that instead of three flat primary colors on the grid, there are seven shimmering Day-Glo pastels. The engineer overlays one grid with a new one of hallucinogenic indigos and magenta and shifts thick pink lines. This is no mythic city he's designing, but the accurate visual simulacrum of a few cells of a semiconductor chip magnified many times. Before CAD it was humanly impossible to design this tiny, smart integrated circuit.

Looking into another screen, you might see a similar image resembling a constructivist artwork, filled with converging lines of force and fragmented geometric objects. The designer touches some buttons and a detail of the picture floods the screen, a close-up of an intricate latticework of pipes, valves, and flanges. He executes another command and a 3-D model, including its entire skeletal superstructure, appears. It looks like a space station, but it is a perfectly scaled model of a petrochemical plant that will be built in an unspecified Arab country.

The programs used to generate these two designs are developed and marketed by companies such as Compeda, of Paramus, New Jersey, a firm that specializes in software used to design and produce both integrated circuits and massive process plants, from chemical to coal, from hydro to nuclear-power plants, and more. Compeda's Plant Design Management System (PDMS) is used by such construction companies as C. E. Lummus, of Bloomfield, New Jersey, to design colossal industrial plants throughout the world. "PDMS allows you to build your plant in the computer before building on site," says Compeda vice-president Carleton Howk. "So when you're constructing it, you're actually building it for the second time."

The 3-D model addresses all problems specific to plant design and construction, including pipework ("the biggest bugaboo"), nozzles, and valves. It provides interference analysis to determine, for instance, whether two pieces of metal occupy the same space, or whether a crucial valve is inaccessible. On-site construction difficulties involving terrain or roads can be anticipated, and every connection and alignment in a project, often many acres in size, can be tested with absolute accuracy. "PDMS is like taking a full-scale plastic model, perfectly accurate, and putting it into the computer," Howk says, "then allowing the design team to get into the computer with a Polaroid camera to take pictures of any part of it from any position."

Traditional hand-drawn blueprint methods account for no less than 25 percent of cost overruns resulting from errors caused by tool limitations. With PDMS, Howk boasts, error is virtually wiped out.

Theoretically, a complete CADCAM setup allows one person to do the following: generate a blueprint on a computer, modify it, give it three-dimensionality. Pictures of standard components of the design can be summoned up at the touch of a button. The computer calculates the sizes of parts and performs analyses to test for wear and tear. The engineer simulates the parts interacting with the whole. He finishes the design and files a geometrically perfect model of it in the computer's memory. Going from CAD to CAM, the computer generates an NC machine-tool path program from the model in the memory. A simulator simulates the tool cuts in the hunks of raw material to verify that the machine will carve out the part accurately. The NC program is then punched onto paper or is stored on magnetic tape. When production time rolls around, the NC program is sent "downline" to the

NC machines by control computers. Machines turn out the product.

Even with today's semicomplete CADCAM systems, there are reports of astonishing productivity gains. In some cases the design/production cycle has been cut in half five years after CADCAM was acquired. There are stories of plant output leaping 50 percent. For example, in 1977 Minster Machine Company, of Minster, Ohio, hooked up a computerized NC flame-cutting machine to a CAD graphics system. This mini-CADCAM setup increased Minster's flame-cutting productivity by 30 percent; its overall weld-shop productivity improved by 17 percent. And today its engineering is on schedule despite a dramatic increase in workload.

For Wall Street, the CADCAM market is a new frontier. The cost of a large CADCAM system begins at about $100,000. There are more than 200 turnkey companies supplying ready-to-use combinations of hardware and software, and the market for their wares has tripled since 1978, with sales breaking through the $1 billion mark early in 1982. By 1986 the market will top $5 billion, forecasts Timothy Gauhan, of Dataquest, an industry-watching subsidiary of the A. C. Nielsen Company. "By 1990," he estimates, "ninety percent of all engineers will work on CAD equipment, irrefutably. And you almost can't have CAD without CAM."

Trying to gain a perspective on CADCAM's "Wild West show," Dick Spann, president of Adage, a graphics company in Billerica, Massachusetts, likens the future of CADCAM advances to the evolution of the calculator—from the 60-pound table models of 30 years ago to today's handheld programmable ones. "Analogous developments are going to occur," he says. "While CADCAM is not in its infancy, it would be folly to call it mature."

The roots of CAD go back to the early Fifties, to military devices that converted radar information into computer-generated pictures. Massachusetts Institute of Technology's giant number cruncher, the Whirlwind, built during World War II as a prototype flight simulator for combat pilots, had CRT displays and oscilloscopes in its control room. But perhaps the most striking achievement in the early days of CAD was the 1963 MIT doctoral thesis submitted by Ivan Sutherland. Entitled *Sketchpad: A Man-Machine Graphical Communication System,* it was a collection of algorithmic programs that was to become the software basis of interactive computer graphics.

In 1970, a year after he founded his own company with David Evans, Sutherland wrote about computer graphics with undiminished awe. "Whereas a microscope enables us to examine the structure of a subminiature world, and a telescope reveals the structure of the universe at large," he stated in *Scientific American,* "a computer display enables us to examine the structure of a man-made mathematical world simulated entirely within an electrical mechanism. I think of a computer display as a window on Alice's Wonderland in which a programmer can depict either objects that obey well-known natural laws or purely imaginary objects that follow laws that he has written into his programs. Through computer displays I have landed an airplane on the deck of a moving carrier. . . . flown in a rocket at nearly the speed of light, and watched a computer reveal its innermost workings.''

The two main graphics systems used in CAD are calliographic and raster. Calliographic creates the familiar wireframe image—the rotating Datsun on TV—as an electron beam moves from place to place on a screen in a pattern that traces out the individual line to make up the picture. Raster displays make pictures the same way as ordinary TV sets do: The image is "painted" in fixed sequence from left to right, top to bottom.

The basic picture element in calliographic systems is the line segment, or vector. To make images move, motions are broken down into "primitive" orders: rotate and scale (shrink or grow), based on mathematics. You can zoom in on a region near an object and the rest of the picture flies outward, too big to be contained on the screen. These outlying parts are removed from the "display list" by another primitive operation called clipping. To create solid objects on raster displays, you model the object's curved surfaces by splitting them into small surface patches represented by equations. For 3-D forms, the curves generalize into polygons, cones, spheres, and even more complex surfaces.

The latest thing in CAD, solid modeling, can build up complex shapes out of previously defined primitive shapes. Alan Barr, a doctoral candidate at Rensselaer Polytechnic Institute (RPI), in Troy, New York, creates vivid peacock-colored shapes on his graphics display. He and his colleagues have developed a system called Superquadratics. It can construct any shape capable of being defined by a quadratic equation. That includes doughnut, star, and dumbbell configurations as well as pastalike ellipsoids and hyperboloids. Where solid objects are to be joined—a nut and bolt screwed together—there are special geometric representations that describe the object as a volume with a smaller volume inside it. To pass through the looking glass, you marry art with mathematics.

There are advanced algorithms for hidden line and surface removal, that is, getting rid of parts of the object that would not be seen if they face away from the viewer, or that become obscured as the object rotates. There are mathematically programmed "intensity values" or shading; they are determined by surface orientation, the direction of the light source, and surface textures. There are algorithms for creating the smoothness of a baby's skin or the roughness of the moon's surface, and even transparency.

The computer is the master of perspective, the art of depth's illusion. The CAD system of Applicon, a turnkey company in Burlington, Massachusetts, has a design package in which the image can be modified from six orthographic views: designers can see it and change it from top and bottom, right and left sides, front and back. The same operation can be done from any axonometric projection, too, rendering it realistic from several perspectives. Applicon's Flying Eye editing lets you walk inside and redesign the interior of your model as well. Its solid-modeling software enables you to determine precise mass properties such as area, volume, weight, center of gravity, and moments of inertia. With "exploded view" you can make a complex gizmo such as an automobile engine fly apart into its smaller components.

With all these fabulous tricks, nevertheless, there is no easy way to describe complex 3-D shapes. "How do you describe a person, a face from hair to eyes to mouth?" asks Mark Fox, head of the Intelligent Systems Lab at Carnegie-Mellon University (CMU), in Pittsburgh. "How do I deal with the extension of my veins here?" He proffers his forearm. "The only person who could model the human body properly was Michelangelo, and even he wouldn't be able to come up with a description that a machine could understand," Fox asserts.

Nonetheless, engineers talk about "before and after CAD" as if it were a religious date. Certainly, by taking the drudgery and routine composition out of engineering, CAD has changed the nature and quality of design. "We can investigate large numbers of alternatives early on in the design sequence," says Steve Fenves, professor of civil engineering at CMU. Instead of having only 3 or 4 mousetrap designs, you can have 3,000 or 4,000. You're bound to come up with a better mousetrap. "And now it's feasible to change a design radically late in the game, too," Fenves adds. "You can have customization of detail. In future designs you will have buildings with much more complexity and the perception of complexity." Architects will work with far subtler concepts, will be able to concentrate on sculptural facets of a building, and will observe how the sunlight hits them.

"Until CAD," says Ken Tull, staff engineer at Lockheed Georgia, "the aerospace engineer was using the same instruments as Euclid and Pythagoras—compass, square, and pencil. With the exception of a radius gauge, there was nothing new. When you put that engineer on a CRT, you give him a whole new world. Instead of having a drawing board three feet by six feet, he's got a board as big as forty-eight football fields.'' He can ''resolution up'' a scale model in a few seconds without losing one centimeter of accuracy.

The most tedious part of design has been analysis—heat, stress, the structural behavior of parts under loads. Software packages such as the National Aeronautics and Space Administration's NASTRAN have been developed to do this work. These analytical programs subject the computer model to various mathematically simulated stresses. The model is broken down into tiny parts called elements, and since there are a finite number of them, the process is called finite-element analysis. Done by hand, it's a mind-numbing chore that generates reams of data. If the analyst did not cover all his bases, he might have missed an important flaw in the design—with dangerous consequences. Linked to CAD, finite-element analysis accelerates the process. The model is often generated in color and lights up brilliantly where stress contours occur.

Mark Shephard, a professor at the Center for Interactive Computer Graphics, at RPI, demonstrates his new system. It is applicable to anything that requires a governing set of rules, from nuclear containment vessels to aircraft. He turns to his CRT, where an image of a sheetmetal bracket glows green. The 2-D object, resembling a flatiron with a hole in it, is traversed with a webbing of little triangles. ''Each triangle has an assumed behavior,'' he explains. "Put enough of them together and you get a reasonably good idea of complex behavior. Historically these meshes were calculated by hand. On this bracket, that would be a several-days chore. On the computer, it takes one hour." He zooms in on an area of the bracket to examine how much load is displaced.

"If I'm concerned with how 'out-of-round' the hole is, we can zoom in on that, or we can 'walk around the bracket.' " He fiddles with the controls. "After you do the analysis, you can go back and make the bracket thicker to avoid stress concentration, or you can take out excess material. When finite-element analysis is made an integral part of the design process, then design is based less on rule of thumb and crude approximations. Although this, too, is an approximate analysis, it verges on perfection."

There is one drawback to near-pefection, however, and that is the tendency to design too close to the line. In the past, engineers didn't have the tools for such fine analysis. So they grossly overdesigned to be on the safe side. At today's close-to-flawless condition, a part is designed to do its thing perfectly, except for the unexpected contingency. "If you push it the wrong way, and it wasn't designed for it," Shephard says, "the gadget will break. The computer allows you to be more efficient. But for the consumer that may be a mixed blessing."

After analysis, the next step is to cut down on production time. "If we can run a graphics program," Fox says, "to stimulate the cutting of a turbine blade instead of running a real half-million-dollar machine and taking it out of production, we've saved fantastically on time." Computervision, the leading turnkey company, recently introduced a "dynamic tool motion" program whereby the engineer at the console can simulate the interactions between any tool and the piece it's working on. (The program is called "tool-path verification.") Computervision's program offers an enhanced reality: The quasi-omnipotent engineer can view the tool's cutting path from any angle at the flick of a joy stick.

In CADCAM's great chain of being things such as tool-path verification constitute middle links that tie CAD to CAM. But how far away is the goal of complete design-to-production automation? "How simple is the job?" Fox returns the question. "If it's a complex three-D shape, nothing goes directly from design to production. But if it's a simpler shape such as a block of metal subject to a milling machine that merely goes up and down cutting holes in it, then it's easy to go from specifications to actual program and production. But systems are becoming much smarter, and more CAM systems can go from increasingly complex shapes to automatic programs—for metal cutting and milling in general."

The ultimate CAM "organism" is the flexible manufacturing cell. The cell concept implies a self-contained series of machines that perform a complete task and are serviced by robots or other forms of manipulators and controlled by minicomputers. The cell feeds itself materials and divests itself of a finished item with minimal hands-on human interference.

Right now there are few complete cells in operation in the United States, and the ones that exist are not very flexible. But in a unique collaboration, Westinghouse and CMU are in the final stages of developing an advanced computer-robot-machine unit that could become the model for future cells. The flexibility is attractive to Westinghouse, which makes more than 250,000 different products. Westinghouse engineers, along with CMU's Paul Wright and his group in the Flexible Manufacturing Laboratory, will install the first cell in a Westinghouse Winston-Salem turbine-blade plant later this year. "Our goal," says Wright, "is to make one batch of turbine-blade preforms [rough shapes] without any human intervention whatsoever."

The Westinghouse/CMU cell is enough to make Hephaestus drool. Its star player is a formidable two-and-a-half-million-dollar, 20-ton, open-die forge called a swaging machine. It sits on a foundation 40 by 60 feet, and its computer-driven hammers beat a piece of stainless steel, called a billet, into an airfoil shape. "It's a twentieth-century village blacksmith," says Gary Shatz, Westinghouse manager of advance manufacturing.

Attended by two robots, the swage operates in an atmosphere where temperatures can reach 2,100° F; this heat is generated by the other main player, a cylindrical furnace. This giant cooking facility heats a billet the size of a large plastic Coke bottle. Placed into the furnace by a robot's arm, the billet is removed by the robot when it's hot enough and taken to the swage to be beaten into a preform. Then still another robot removes the preform from the swage and hauls it to a vision system for fully automated inspection.

If the preform is defective, Wright explains, machines correct machines. A diagnostic fault-detection system not only finds imperfections and breakdowns but also has built-in strategies to accommodate and correct them. These decisions are made in the software.

Wright has his own simile that explains the cell. "These machines are like individual musicians in a chamber orchestra. They need a conductor to make sure they're all playing in sequence. When does one tool start up, another stop? How many blades does one machine make before it switches over to another task?" Programming authority is divided into a local André Previn conductor code that resides in a supervisory computer (typically a PDP 11) and a Beethoven composer code in a large mainframe computer such as VAX 11/750. There are myriad programming problems; besides the vision systems and the fault-detection codes, there is even a robot-avoidance program to keep the automatons from bashing into one another. "We have to describe forbidden places where they can't go."

José Isasi, Westinghouse manager of manufacturing planning at the Winston-Salem plant, is happy about his cell. "This preforming is a high-energy process," he says. "We're talking about extreme high temperature, huge equipment never before used in a cell, things that put out tremendous forces. Usually there are a lot of people to watch them. In our case the proof of the pudding is gonna be that we turn off the lights and go home and let the computers run it all night. And when we come back, and if the cell is still there and it has not blown up, then we'll know we succeeded. That's the test.

We're gonna depart the area and let the computers alone with the robots and sensors and everything." When Isasi gets going, he sounds like CADCAM's Father Guido Sarducci.

"And the computers will have to run several kinds of equipment, all on the brink of going wrong," Isasi continues. "And you're talking about a lot of heat floating around, but there's a lot of computer horsepower on every level so that we can detect if something is about to happen before it explodes or blows up or destroys the two-and-a-half-million-dollar machines!

"It's a hostile environment," he points out, "and just because it's such an aggressive environment, the eye of the robot is protected by a fancy enclosure. It keeps the lens clean and air-conditioned. We actually had to build a computer room right on the factory floor to house all the computer horsepower. Some people call it Mission Control." Westinghouse is installing a closed-circuit TV system "so that in the future," Isasi says, "we can have one or two guys on a central control overseeing the entire factory."

"Physically it would be possible to run the cell from CMU in Pittsburgh," adds Jerry Colyer, project engineer at Winston-Salem. Later this month Operation Warm Billet will execute a sequence from beginning to end—billet in, preform out. It will be the first time a supervisory computer code and diagnostic fault system, plus robot vision, will be used in a U.S. heavy manufacturing situation. They plan to produce one preform per minute.

How will cells be used in the totally automated factory of the future? "First you've got to understand how each cell could influence all the others,'' Wright muses. "You'll need many more sensors. Each cell will have to be coordinated with the next." And robots will have to do all the running around. "Now few factories are laid out with robotic transport in mind," Wright says.

If you were going to invest in a CADCAM setup, where would you start?

On today's market you can buy CADCAM systems of varying degrees of integration—from computerized NC machines to top-to-bottom design/production units. We looked at a few of them. At the Danetics/Summit plant in Bozeman, Montana, a spokeswoman named Kathy Deeter sang praises of her company's wares. One device called the Bandit is an automated contrivance for lathes, grinders, and chuckers. You can program it as you would a computer to cut out any pattern on any product on up to three axes. It can indent surfaces and can nearly carve out a rounded ball. "And we're coming out with Bandit Three," Kathy enthused. "It's just superduper. Everything you ever wanted—graphics, a CRT that shows you the object as it would look cut out, all sorts of look-ahead devices and human engineering." *Human engineering* is CADCAMese for making the programs easier for morons to use. *User friendly* is a synonym.

Deeter also described the Quick Draw tool changer, a vulturous-looking machine that actually goes under the appellation of "dedicated" robot: It is designed for only one application. The Quick Draw's automated arm reaches into a tool crib, picks up the right implement, screws it into a little claw, makes the cut, takes the tool out of the claw, and puts it back into the crib. Then it goes on to the next tool and task. Danetics also offers the Intelligent Driver, a smaller version of the Bandit, and the Task Master, a medium-sized Bandit.

"Somebody bought twenty Task Masters the other day," Kathy told us. "They make bra straps with them. The Task Master electronically picks up the strap, measures off the right amount, brings it over to the laser, which lases the strap onto the bra. No stitching involved." The Bandit people also offer retrofitting. A Bandit can be attached to an old lathe, mill, or grinder, "so instead of a shop person standing there, you've got a Bandit." Although it declines to reveal its annual sales, Danetics says it is selling "hundreds of thousands" of machines through a distributor to markets in the United States, Europe, and Japan.

Computool, of Minneapolis, boasts of being the only thoroughly computerized turnkey CADCAM system available. Computool specializes in the production of precision tooling (making things with tools), plastic injection molding, and die casting and forging. Computool's software permits programming at the Compuscope whereby a designer can vary the size of his tool and the rate the tool does its job, and he can contour the workpiece in 3-D right at the console. "We're having problems handling the number of people interested in the system," says Computool's understated president, Fred Zimmerman.

Zimmerman says one customer made $6 million worth of molds on a Computool system. Another had 122 toolmakers and now has 30. "And they're doing twice as much tooling." He reminds us that, even though metalworking is the segment of industry utilizing CADCAM to its fullest advantage, 95 percent of tooling is still done manually. "But CADCAM's beginning to pick up speed," he reveals. "Our system is effective for sixty percent of applications. Yet you can't just sit down at a computer and type *mold* and expect one to come flopping out straightaway. It's extremely fast at what it can handle."

What Computool-type CADCAM can handle most effectively is a job where there's an element of repetition. "If a computer is doing multicavity molds, or making families—such as families of mixing bowls—where there are size considerations, or a logo that appears on many parts and is actually molded into those parts," he explains, "then you get great economy." Like many CADCAM companies, Computool makes no attempt to close the loop with robotics. "We don't know beans from buckshot about robotics. We drive milling machines to cut hot and cold steel, and there are enough problems in that," Zimmerman concludes.

From the design of mammoth petrochemical plants to nuts and bolts, CADCAM's interpenetration of industry promises to alter everything, including the life of the white-collar worker. What will CADCAM do to the manager of the twenty-first century? we ask CMU's Mark Fox.

"Eventually it will eliminate him," he replies with a boldness he then attempts to qualify. In the twentieth century, he backtracks, integrated computerized systems will actually allow the manager to keep his job. Fox characterizes the modern corporate man as one who is "floundering in a morass of complexity." Within the conglomerate, white-collars are running around like headless chickens while upper management is busily putting out fires everywhere. "The manager has to deal with everything on an emergency basis," Fox says. "There's little time to think, no time for long-range planning. People whose minds can encompass all that's happening can't handle it, because they don't have access to the right information."

Fox's solution to this chaos is the Intelligent Management System, which he and his group at CMU are developing. Its immediate goal is to aid managers with the avalanche of data coming at them continuously. But its long-range goal is to automate the management of whole organizations, functions such as scheduling, planning, simulations —things that white-collars think about.

"We're working toward the factory in which there are no managers." He grins gleefully. "Just this distributive intelligent system. There'll be microprocessors on every floor; processors corresponding to accountants and programmers—all sorts of processors all communicating, trying to achieve perfection."

How did Fox arrive at this vision of Utopo-factory? "It has been observed," he says, "that in smaller batch-size factories, white-collar labor accounts for a large quantity of the total labor cost, in some cases exceeding fifty percent." Fox cites the Japanese Hitachi Company, which in 1972 found that 100 white-collar workers "touched" or interacted with one product. Hitachi automated scheduling and reduced the number of white-collars who touched the item to 30. In existing U.S. factories,

Fox says, "the product or part is often lying there on the floor eighty percent of the time, waiting for something to happen. Even if you get CADCAM to increase the productivity of your factory machines, you won't have much improvement until you reorganize your business radically.

"CADCAM is a big laugh, from my point of view," he continues. "We're fooling ourselves if we think robotics is going to be the savior of industry. When you're talking about service industries where a lot of production is job-shop-oriented, then *management* is the bottleneck. But we don't see the Intelligent Management System replacing people for a long time," he says.

Others agree that CADCAM won't take over factories overnight. "CADCAM is a revolution of such enormous scope," says Bela Gold, director of the research program in industrial economics at Case Western Reserve University, in Cleveland, "that most people are unaware of where it's leading. But it's going toward a totally different way of organizing manufacturing processes and a totally different perception of how those processes function." Gold thinks the traditional U.S. corporate structure is addressing the CADCAM revolution badly, if at all, and contrasts the American structure with the Japanese "top-down" organizational system. "The Japanese implement one component at a time," Gold says, "but always in terms of how that component fits into the overall architecture. You've got to have tight conformity of successive stages. Otherwise the programs don't work, and the machines don't match each other or the programs."

According to Gold, the worst impediment to this

revolution is top management. "They get their advice from the V.P. of engineering or manufacture, and many of these old guys have been out of it for twenty years, and, by the way, I'm an old guy myself," the sixty-seven-year-old economist laughs. "They reinforce the innate caution of a lot of senior executives and say, 'Wait till you can press a button and get a custom-tailored suit.'

"You can't just plug these things in," Gold continues. "You've got to keep adapting them to the changing pressure of the production and marketing situation. And if you don't have a staff trained for it, the tools will be underutilized or will break down. And in robotics," he adds, "they are trying to replace people in operations that already exist, instead of redesigning the production in accordance with robots' capabilities. We are adapting robots piece by piece, and they are not going to fit together. Likewise with CADCAM systems."

With blue-collar unemployment already at 12.5 percent, there is no way to avoid CADCAM's most looming question: Will the automation revolution precipitate a pinkslip blizzard? When asked about massive dislocation of the workforce, most pro-CADCAM people voice the same reassuring argument. "There is no problem," Zimmerman maintains. "The average toolmaker in the United States is fifty-six. Ten thousand a year are retiring, and three thousand a year are coming in. We like toolmakers and find they make excellent operators for our system. We're not trying to do away with them, only to use them more productively."

"Layoff the big fear!" exclaims Westinghouse manager of communications Peter Ryckman. "We haven't laid off a single worker because of CAD-CAM. But the pool of eighteen-year-olds decreased by thirty percent in the last four years. There's a strong feeling that any company not involved in programmable automation because of manpower resistance will have to do it down the line."

Tom Moser, manager of engineering systems for Westinghouse, tells this joke: "Harris, you've been replaced by one computer, three programmers, four field engineers, and two systems analysts."

And CMU's Paul Wright doesn't see CADCAM development as all that different from the rise of modern banking and telephone industries. "Phone employees used to occupy themselves at switchboards. Machines do that today, yet Ma Bell employs just as many people. My scenario is that far fewer people will be involved in grubby tasks of the Dickensian sort. Those same people will be retrained and diversified into programming tasks, and of course there'll be a huge need for repair and maintenance."

Not that there won't be dramatic changes in the nature of work. It is highly conceivable that by the year 2025 programmable automation will have replaced most operative jobs in manufacturing (about 8 percent of today's workforce) as well as a number of nonmanufacturing jobs, CMU's Robert Ayres and Steve Miller say in their paper "The Impact of Industrial Robots." "Concerted effort should be made by the public and private sectors to redirect the future workforce to these changes," they advise.

In the long-distant future, the descendants of CADCAM technology may bring about the end of work as it has been experienced ever since the first humans picked up a stick and scratched the soil. We may not need money eventually, nor any other form of barter system. Instead we may spend our days chasing enlightenment or sitting around enjoying our piña coladas while computers and robots program and manufacture our durables and scuttle about busily building more of themselves, preparing for that next evolutionary step people such as Robert Jastrow talk about—the rise of silicon intelligence.

TO PAY FOR
THE FUTURE

By James S. Albus

They're coming. But we're not ready. In our factories and in our homes. In stores, offices, and the places where we play. Doing all our work, liberating us from drudgery, creating new goods at a rate and of a quality unknown in history.

Robots. Not the happy tin men of science fiction, with snappy patter and red lights for eyes. Not R2D2 and C3PO. These real robots will be sophisticated working machines, programmed to execute specific tasks extraordinarily well. They will see a little, sense a little, move a little, and "think" only enough to carry out their assigned tasks.

The robots will be superb information-handling machines that give computers the manual dexterity to perform such tasks as welding in an assembly line and selecting a stock portfolio. From a base in the manufacturing industries, robots should spread throughout our society, eventually taking on every kind of job and creating new ones.

These ultimate, responsive working machines will transform our economy and our lives the way the discovery of steam power revolutionized the world 200 years ago. Steam unleashed a vast supply of raw power that changed industrial and commercial enterprises. Robotics will create an entirely new form of work. Steam took over jobs that required lifting, shaping, moving, and heating. Robots will supplant human workers in tasks that require manipulation, measurement, complex interaction, and quality control.

Our modern age resulted from that first industrial revolution. The future belongs to the second industrial revolution, the robotic revolution.

Yet we are not prepared, in the slightest degree, for the changes that loom ahead. It would be nice to paint a rosy picture of happy workers and happy robots building a better world together. But real life usually doesn't work this way.

The first industrial revolution, though better than what had gone before, propagated untold misery in the lives of workers. The world's lack of preparation for the changes that occurred led to sweatshops, labor riots, snarled political allegiances, and enormous suffering. Karl Marx was moved to call for a revolt by the world's oppressed workers as the only means to rectify the wrongs the industrial revolution had brought with it.

The second industrial revolution will have an even greater impact on the world than its predecessor did. And our social, economic, and political fabric is simply not strong enough now to withstand the stresses robot technology will introduce. If we don't change our present patterns, the inevitable and widespread use of robots will come at the cost of high unemployment, high inflation, social unrest, and violence both physical and psychological.

We face the cruel paradox that our greatest hope for an idyllic future is also the greatest threat to our society's orderly existence. If we ignore robotics, we will be outdistanced in the world market by the forward-looking countries that adopt it; their products will be both far cheaper and better made than our own. If we move into the Robot Age in a haphazard, unplanned manner, we will wrench our society apart within a single generation.

So the only sensible course available to us is extensive and early planning for a smooth transition from today's world of human workers to tomorrow's world of robot workers. We can avoid the pitfalls of the second industrial revolution if we begin working now. We can democratically create programs now that will be far better than those that might be imposed upon us for the sake of survival.

Consider the magnitude of the change that approaches. Today most of us are inextricably bound to our jobs and wages. We spend most of our waking hours on the job or in transit to and from our working place. Our livelihood depends upon the money we receive in exchange for making a product or performing a service.

In a robotic society we won't work. The robots will do most jobs more efficiently than we could do them. Such robots can usher in an age of superabundant, very inexpensive goods. But there's a hitch. If humans don't work, do they deserve to be paid? Where can they get the money to buy all the wonderful and inexpensive products the robots will produce?

We can't have a society of unemployed people who can't afford the products made for them by the robots that took away their jobs. That's absurd. We must devise a scheme that will shift the manner in which we receive income without endangering either our standard of living or our self-respect.

The first challenge of robotics is: How do we pay for the robots? Vast sums will be required for investment if we are to benefit from robots. At first, worker machines will be quite expensive; they will have to be constructed by people. Eventually one generation of robots can be programmed to build the next, bringing the cost of robots steadily downward. But where will we get the money for the all-important first push into the robotic age?

One way or another, the money will have to come from the people. In the past, when important challenges required huge investments, the government went to the people and asked them to invest in public bonds. During World War II, for instance, War Bonds represented a willing investment by the people in winning the war.

Robotics is the challenge of the future. The government could move us to accept it by forming a quasi-public agency to sell Victory Bonds for the Future. The money-gathering agency that issues the bonds might be called the National Mutual Fund (NMF). Here is how it might work:

First, a massive campaign of advertising, promotion, and speeches would be undertaken to explain to the American public why robotics is so vital for their future. Without it, they would be told, the next generation of Americans will live in a second-rate society whose goods won't sell on world markets. Jobs will be scarcer than ever. The dollar will be so depreciated that it won't buy a stick of chewing gum. Failure to accept the robotic challenge will put the United States in an ever-deeper fiscal hole.

Americans will be encouraged to invest as much as they can—until it hurts—in Victory Bonds for the Future. To stimulate public acceptance further, the bonds could pay interest pegged to the cost of living. Investors would always receive, say, 5 percent interest more than the year's average rate of inflation.

With the money it receives from public investors, along with money appropriated by Congress or gathered from other government sources, the NMF would build a significant supply of cash. The NMF can use this cash to finance companies that want to adopt robot technologies. The firms can issue special stock offers, which the NMF can buy. They can then use the money from the stock sales to install robots.

Now let's look at the other side of robot economics: As the new technology expands, workers will be pushed out of jobs. We must plan to ease their discomfort and make up for their financial losses.

One way would be to allow the workers to own the robots that replace them. As owners, they could lease the robots back to their former employers for use at their old jobs. This plan is highly speculative, of course; it would require extraordinary cooperation among workers, unions, the owners of businesses, and government.

Another method might be to establish huge employee stock-option plans. If employees are given large blocks of stock in their own company, they'll benefit directly from the increased productivity the robots will furnish. The dividends on their shares could offset their lost wages. This plan would ease the burden only for the labor force now employed; it would do little for those who enter the labor force after the transition hits its stride.

Also, the fortunes of displaced workers would be tied to the profitability of a single company. In a society in which only half or one quarter of the work force remains employed, the others displaced by robots, a bankruptcy in one company would be disastrous. The dividends would disappear, and no jobs would be available to offset the loss. The government might have to perform Chrysler-like bailouts on a herculean scale to maintain robot companies.

A third course, probably the most attractive one, would be to use the NMF as a conduit for payments to displaced workers. A stipend from the dividends the NMF receives on the stocks it buys could be guaranteed to each unemployed human worker. Every time the NMF finances a company into the Robot Age, it will get stock, which pays dividends. As robots spread through the economy, the NMF will hold stock in more and more companies, receiving greater and greater dividends.

As more workers are put out of jobs by robots, the NMF will have steadily more money to assist them. It will work somewhat the way Social Security or unemployment compensation does now, but the money distributed will not be tax revenues; it will be dividends, like those of any mutual fund.

Not all workers who lose jobs to robots will be put out of work. The adoption of robotics, in its early stages, will eliminate many jobs, but it will open up many others. As assembly-line jobs disappear, construction of new robot factories will require skilled hands. As technical services go robotic, more programmers will be needed.

The shift from a worker economy to a robot economy will be gradual, spanning at least a generation. If the basic institutions and plans are ready at the inception, there will be time during the process to adjust to the changing situation. The shift to robotics could be managed almost on an industry-by-industry schedule, controlled and reasonable, if there are such agencies as the NMF ready to oversee the change.

One other problem our robotic control plans will have to cope with is inflation. As companies and the government pour money into robotics, the economy could become dangerously overheated. For a time the relationship between wages and prices could go badly off balance. Eventually the efficiency of robot workers will bring the prices of all goods well below current levels (raising the quality, too). During the interim, however, fluctuations in the economy might be dramatic.

This might necessitate establishment of a monetary control system above and beyond the Federal Reserve System. One possibility would be an "enforced savings" program. Whenever prices begin to gallop away, the government could impose a special deduction on all paychecks to draw money out of the economy and thus lower demand. This money would be designated enforced savings, placed in special savings accounts where it would stay—gaining interest—until the economy cooled.

Obviously, such a plan would represent a serious sacrifice for workers. It would also be an unusual infringement upon the rights of American citizens. But if we want to adopt the technology we need for the future, a few such exceptional efforts will surely be required. The alternatives would be ballooning inflation and an uncontrollable economy that could ruin all plans and throw the United States into financial chaos.

Even if our entry into the future goes smoothly, robots will not bring utopia. No machine can, because the final barrier to human happiness lies in human nature.

Robots will bring in an age of universal prosperity. They will help us stretch out natural resources and create synthetic resources when the natural ones are exhausted. They will improve the quality and availability of all goods. They will help us husband our food supply to provide an adequate diet for all humankind. They will encourage us to reach out and explore the universe, the earth, and ourselves.

But the robots won't solve our paramount problem: overpopulation. Our unchecked growth has outstripped the productivity of every technology ever devised and sucked dry every pocket of useful resources ever discovered. Our exorbitant numbers have strained world industries to the limit.

Robotics will give us one last chance to overcome this fundamental bane of civilization. It won't solve the problem, but it will give us time to solve it ourselves.

Perhaps this is the best use to which we can put the seamless stretch of work-free time we'll have. Can we control ourselves? Can we create a social order that keeps within the limits of its resources, space, and food supply? Can we make an orderly transition to the frontier of the stars before our expanding population impoverishes even the robotic economy on Earth?

Freed from arduous toil, awash in material goods, we will have to become a new type of human. Will we use the freedom of this future age to secure mankind's stake in the universe, or will we fritter away this opportunity in boredom and dissipation?

Robots pose this ultimate question to us, but only we can answer. Which will it be: an orchestrated future, or the last dance?

PART SEVEN:
CYBERCONFLICTS

PROGRAMS
FOR PLUNDER

By Roger Rapoport

If I had been a traditional embezzler,'' says Malcolm Senn, ''they probably would have arrested me within six months. But with the computer I never had to touch the cashbox. Being controller made it easy. First I set up fourteen phony suppliers. Then I programmed the unit to pay automatically for nonexistent goods and services from my dummy companies. That way I could be skiing in the Alps while my California employer mailed checks to these fronts.

''The system earned me a million dollars over six years. By then I was ready to quit and start enjoying the houses, planes, and boats I'd accumulated. I knew if I just left, my successor eventually would figure out what I'd done. Since I can't stand suspense, I began leaving clues to help the auditors catch me. I wanted to have my day in court, serve a brief sentence at a country-club prison, and then live off the money I'd stashed in Switzerland. But the auditors were impossibly slow. Finally, in desperation, I started bouncing checks. It still took them another three months to catch up with me.''

Like most of the 800 other people whose computer crimes have been uncovered in recent years, Senn set up his automated theft with no previous lawbreaking experience. Until he began bilking his employer, Senn had never shoplifted so much as a chocolate bar.

Sitting in the living room of his rented Mediterranean-style home on the California coast, he discusses his lucrative data-processing heist with the insight of a business administration professor. On the coffee table is the latest issue of *Fortune;* in his right hand is a frosty glass of white wine that sloshes over the rim every time he gestures to emphasize his pearls of wisdom. Thanks to his pseudonym, Senn will have a talk tomorrow morning with Big Board companies anxious to secure their electronic data-processing (EDP) facilities.

The prospect of hiring a San Quentin alumnus to give advice on fighting computer crime might not appeal to every corporation. However, after losing over $2.1 billion to electronic thieves during the past 20 years, industry waits in line for the services of men like Senn.

And these documented electronic frauds represent only about 15 percent of the total, because 85 percent of such incidents are never reported to police. "Companies are afraid publicity will stimulate more ripoffs like this," Senn explains. "In many cases they prefer to quietly discharge larcenous employees. Some even receive severance.''

More than ever before, computer crime pays. While the average American embezzler takes home $19,000, computer bank frauds average $450,000. Safer than armed robbery and quicker than a mugging, electronic heists can usually be accomplished during banker's hours. Just a few minutes at the right terminal can sometimes net a year's income. Unlike mobsters, data-processing cons don't have to bother with smuggling, shakedowns, or enforcers. There are no middlemen, burns, or deadbeats. And law-enforcement agencies frequently don't understand fully the complexities of EDP fraud, and they can't crack it.

Since 1958, when a computer caught a stockbroker working for New York's Walston & Co. who had stolen $277,607, no industry or agency of government owning EDP equipment has escaped the attention of high-tech con artists. Biscuit manufacturers, insurance companies, hospitals, colleges, and the U.S. Army have all been victimized. One disillusioned employee at a San Jose, California, computerized billing firm decided to get back at his colleagues by walking out with the entire program inventory. Caught without spare copies of missing bills, the marketing manager committed suicide and the company filed for bankruptcy.

Who are the EDP swindlers? Donn Parker, a consultant with SRI International who is considered the Sherlock Holmes of computer crime, has met more of them than anyone else. Sitting beside a five-inch-thick printout he maintains on electronic theft at his office in Menlo Park, California, he says, "Most perpetrators are young, eighteen to thirty years old. They tend to be amateurs with no prior criminal record. They love to play computer games and are fascinated by the challenge of trying to beat the system."

Many of the convicted thieves Parker has met began their life of crime after taking college computer-science courses. "Teachers like to show students how to crash the computer," Parker says. "Ostensibly this is done to familiarize them with the system's inner workings. But things inevitably get out of hand. A couple of years ago eight seniors at my son's high school broke into the district unit and gave all their classmates straight-A report cards. The seniors at a rival school across town got nothing but F's.

"Instead of discouraging this sort of behavior, many colleges let their students form crash clubs. They actually compete to discover better ways to compromise equipment. The result is that universities are turning out a whole new generation of computer criminals."

Institutions hit include Stanford, the University of Toronto, Georgia Tech, and the University of Wisconsin. At Queens College, in New York City, a student moved up the grade point averages of a dozen friends by tinkering with the school's IBM computer. He also improved enough of his own marks to win a Phi Beta Kappa key.

Twenty-six Caltech students once used the campus EDP unit to print out 1.2 million McDonald's contest entry blanks by using a simple Fortran program. Since they had a third of the entries, it was no surprise when the group won a Datsun station wagon, a year's free groceries, and a slew of $5 gift certificates. McDonald's, chagrined by the whole

312

affair, staged another drawing, excluding computer-written entries. Executives of Burger King were so delighted that they set up a $3,000 Caltech scholarship in the name of the senior who orchestrated the winning effort.

Even without crash clubs, schools breed "hackers," who spend most of their working lives close to data terminals. "Wherever computer centers have become established," says MIT Professor Joseph Weizenbaum, "disheveled young men, with sunken, glowing eyes, can be seen sitting at computer consoles, their arms tensed and waiting to fire their fingers, already poised to strike at the keys, their attention riveted like a gambler's on the rolling dice. They work until they nearly drop, twenty, thirty hours at a time. Their food, if they can arrange it, is brought to them: coffee, Cokes, sandwiches. If possible, they sleep on cots near the computer. But only for a few hours. Then back to the console or the printouts."

What makes these hackers good criminals is their ability to exploit automation. As machines replace humans in such routine business activities as accounting, payroll, ordering, and shipping, new ripoff opportunities develop. Access to EDP units is easy. "Give a bright fifteen-year-old computer buff seven hundred fifty dollars, send him down to Radio Shack, and he can create a system that will get him into just about any company computer in town," says Oakland electronic-security consultant Robert Abbott. "It may take him a little while, but he'll do it."

And once these bright young fellows obtain access, they can compete with the multinationals.

Jerry Schneider, while working his way through UCLA with a small intercom-installation firm named Los Angeles Telephone and Telegraph Company, succumbed to the lure of a quick strike. Eager to start selling communications equipment, but hopelessly undercapitalized, he studied the phone company's computerized ordering system and, posing as a Bell worker, obtained all the codes he needed for ordering supplies.

For six months Schneider regularly ordered phone equipment, which he resold to customers of LAT&T. By offering some of the best prices in town on these hot items, he attracted big business. Even Western Electric's local office went to him when it ran short of a particular control unit. Schneider promptly sold Western Electric a hot piece of its own equipment, without the identifying marks.

Only after stealing $1 million worth of phone equipment was Schneider turned in by an employee who had been denied a raise. The judge sentenced Schneider to 60 days at a Malibu work camp; the phone company seized $100,000 worth of equipment and settled a $250,000 lawsuit for $8,500. Expenses, legal fees, and bad debts left Schneider with only $42,000 in his account. Unfortunately, by the time he got out of jail an accountant had disappeared with the entire sum.

Schneider decided to rehabilitate himself by opening a security consulting service in Beverly Hills, California. Sitting in the lobby of the Century Plaza Hotel, he says, ''Even after I had been caught, one of the phone company's employees was using a method similar to mine. The company is still doing things the same old way, although the system is a little more secure physically. I'd suggest they check employee-ordering codes to see whether they're being abused, stop using a line anyone can call in on, and change the computer's phone number every week. Hell, they're still using the same number that I used to call."

Schneider's theft was one of the more sophisticated assaults on a computer. Data-processing skill, however, is not a prerequisite for success. Many outsiders have scored with just a few days' work. One man who helped himself to deposit slips at bank tables in New York, Washington, and Boston put his own account number on the bottom of the slips and replaced them on the tables. Hundreds of people unwittingly contributed to his account. He disappeared after cleaning $750,000 out of the three banks. And the police still haven't been able to find him.

A common technique is to recruit the employees who control million-dollar EDP units but earn only modest salaries. In one of the better efforts, a South Korean gang penetrated the U.S. Army's inventory-control computer at Taegu. They stole more than $17 million worth of food, uniforms, car parts, bulldozer track, gasoline, and other commodities and fenced their loot to Korean contractors, soldiers, and politicians.

At a Pompano Beach, Florida, harness track, clerks working with bettors programmed the computer to accept conditional wagers. On losing bets, the employees simply canceled the bet. On winners, they printed up the valuable ticket. The system netted two professional gamblers $90,000 before it was broken by racing officials, who revoked the licenses of three employees implicated in the swindle.

It took Florida authorities considerably more time to quash a similar operation at Flagler dog track. Six track employees made over $1 million from a five-year trifecta scheme. After each race they held up calculation of the winning values and punched out extra tickets for the right combination. Then they fixed the computer records that showed the actual number of winning bets. An obliging friend cashed in their tickets.

Another gambler, Roswell Steffen, embezzled $1.5 million from his employers at Union Dime Savings Bank, in New York City, to support his $30,000-a-day life-style. "I would go through the computer tapes at the end of the day and see whether any new large-balance accounts were open," he recalls. "Then I would use the system's override and make a correction for about half of the account's balance—fifty thousand dollars, for example—and use that money for gambling."

Whenever a discrepancy was uncovered, Steffen adds, "I would fake a call to the data-processing department and reassure the teller it was a simple error, which I could correct. Then I would have to use the correction feature to take fifty thousand dollars from another account and deposit it in the first one."

Steffen finished far ahead of colleagues who limited themselves to occasional dips into customers' Christmas Club accounts. But the $275-a-month employee was undone when a raid on a bookie's "boiler room" revealed his betting slips. Steffen was sentenced to 20 months.

While some experts believe screening can weed out risks like Steffen, IBM's principal architect of computer security disagrees. "I don't believe personal integrity is a continuing characteristic of an individual," says digital-equipment manufacturer Robert Courtney. "We're all subject to temptation, and we can't gauge what someone might do under stress. Suppose that you hire a person because you're impressed by his integrity. Then his mother-in-law needs an operation she can't afford. His sense of personal responsibility induces this otherwise honest worker to embezzle. So the very thing that led you to hire him is what prompts him to rip you off."

Sometimes even experts like Donn Parker, who has interviewed digital cons from San Quentin to Rikers Island, find it hard to spot a potential thief: "In this business you never really know whom you're dealing with," Parker says. "Not long ago I was on a Los Angeles computer-security panel with a local consultant named Stanley Mark Rifkin. A few months later I picked up a newspaper and read that he had stolen over ten million dollars from Security Pacific Bank."

Rifkin, a thirty-two-year-old computer consultant, worked for a firm servicing the EDP unit at the bank's downtown headquarters. After learning the secret financial-wire-transfer code, Rifkin called Security Pacific and moved $10.2 million to his account in New York City. Then he promptly shifted the funds to a Zurich account, flew to Switzerland, and purchased $8.1 million worth of diamonds from Russalmaz, the Soviet state diamond agency.

Because confirmation lags behind transfer orders on the federally supervised financial wire, it took the bank nearly a week to discover the theft. Rifkin was arrested five days later at an apartment in Carlsbad, California. He had on him $12,000 in cash from one sale he'd negotiated with a Rochester, New York, jeweler and 19 pounds of diamonds worth $13 million at retail. He pled guilty to two counts of wire fraud and was sentenced to eight years in a penitentiary.

Why didn't Security Pacific insist on extra confirmation when Rifkin asked for such a large transfer? "You don't have a signature, because wire-transfer systems aren't capable of signing things," a security officer explained. "So you use the code. Many transfers originate by telephone, and if the man calling has the right personal identification code and the right daily code, you automatically transfer the funds."

Even armed guards, passkeys, and secret codes can't prevent some executives from using their computer to victimize customers. In the $2 billion Equity Funding case, believed to be the largest known automated fraud, high-speed computers spit out fictitious insurance policies.

"The computer was the key to the fraud," California Insurance Commissioner Gleeson Payne explained. "Under the old, hard-copy methods of keeping insurance records, you couldn't build up bogus policies in this kind of volume."

The FBI has trained more than 400 agents to combat these cybernetic felons. Most of these agents have accounting backgrounds, necessary to gather evidence. But some recent investigations have shown the difficulties of putting together a case in the computer-crime field.

A further complication is that today's laws don't adequately define EDP crime. Since computer time itself is a commodity, any programmer who uses the system to print up a Snoopy calendar is technically dipping into corporate assets.

Over drinks at a restaurant in the technological heartland of California's Santa Clara Valley, a programmer sketches some of the problems: "Several years back, my computer-service company was going after a competitor's contract with a Sacramento aerospace manufacturer. I worked up a program that was equal in all respects to the one provided by their vendor. But my boss wanted to be sure our offering had everything the customer was already getting. The potential customer had already given us a copy of his existing program. Unfortunately, a colleague had locked it in his desk and left for the day. In a hurry, I called our competitor's computer and had it print out all the data that my boss said he wanted to see.

"Several weeks later the competitor filed a complaint against me for computer theft by telephone. I had to go through a long trial, which ended with a three-hundred-thousand-dollar judgment against me for stealing something we already had. Eventually we were able to settle with the competitor by trading some computer equipment with them.

"But what was the point of going through all that hassle and expense when I didn't do anything more dishonest than checking a book out of the library? Also, we didn't get the aerospace contract.

"What I'm afraid of is that these laws are going

to be turned against people like me who take nothing of value. We're easy targets, because technical violations, such as using a few minutes of computer time for personal business, are inevitable.

"But there are wide-open opportunities being created by turning the computer into a common carrier, and they are not getting enough attention. As more and more EDP units do message switching, the difference between IBM and ITT is disappearing. And with time-sharing around the world it's easy to plug into someone else's program.

"To ruin a competitor, there's no need to steal. A vindictive employee could cause enormous havoc by erasing data. I know one guy who had a program destroy itself on April 1 as a joke. Instead of designing safer systems, they're just going to make the programmer the scapegoat."

John Taber, an IBM programmer, has similar worries. "Executing a search warrant at a major computer center would be a nightmare," he says. "Just looking for evidence of fraud at my place in San Jose requires running a large computer round the clock for a week. And while they are going through thirteen thousand tapes, we wouldn't be able to get any work done. It's the sort of disruption that can force some companies into bankruptcy."

Senator Abraham Ribicoff, of Connecticut, last year sponsored a computer-crime bill that would have imposed sentences of up to five years for EDP felonies, but the proposal died in committee. With the legislator's retirement, it is not certain when—or even whether—a new bill will be drafted and introduced in Congress.

Courtney believes new laws may not even be necessary, however. Improved system-control programs, restricting users to data they legitimately require, and cryptographic devices might provide adequate protection, he believes.

"Internal auditing has also improved, thanks to the new Foreign Corrupt Practices Act," Courtney points out. "This law, which resulted from the international corporate bribery scandals of the past few years, puts everyone on notice that he or she is responsible for his or her own acts. More detailed accounting standards make it virtually impossible to transfer assets into secret funds."

But automation has created irresistible temptations. In Wheaton, Illinois, a policeman peddled confidential criminal files to a trucking company that was reviewing job applicants. After he had accepted a $14,000 bribe, he was indicted by a grand jury on 24 counts of official misconduct, conspiracy, and theft.

In other countries computerized cash machines have helped change the *modus operandi* of kidnappers. Instead of asking his victim's parents to drop off the ransom at a remote location, Masatoshi Tashiro instructed them to put the money in his account at the Dai-ichi Kangyo Bank. Police didn't know which of 348 remote teller machines he would use for the first withdrawal. They finally decided to send detectives to wait near each of the bank's machines. When Tashiro inserted his magnetic card into a unit at Tokyo Station, the computer immediately flashed his location. When the kidnapper was arrested, he was holding some 290,000 yen.

"Don't let that kind of story deceive you," Senn admonishes. "The guy who wants a quick hit has problems. But for someone who's willing to move slowly and systematically, as I did, times have never been better. I see the opportunities every day at the EDP firms I consult for. Once you've established a workable system, everything else is just details.''

Senn stole from his employer after failing to receive a promised bonus. So much money was flowing in and out through his dummy suppliers that it was impossible for his superiors to catch on. "I ran the data-processing firm that serviced our outfit, besides being controller. Printing up phony books after work for the auditors cut into my sleep. But with my bogus outfits and inflating costs on legitimate acquisitions, the money was mine.''

By day Senn was just another diligent officer putting in 50- to 60-hour weeks. But at night and on weekends he was out tinkering with planes that took him to vacation homes in Arizona and Mexico. "I'd just put the autopilot on and fall asleep. When the vector changed near Palmdale, there'd be a beeping sound that woke me up. Much of the money went into Swiss gold, which has turned out to be a wise investment. After the authorities caught up with me, all they could attach were the planes and the houses. My company promised to drop all charges if I would simply return the money. I insisted I didn't have it.''

One reason Senn lied about it was that his lawyer had promised him that plea bargaining would limit his jail term to no more than a year and a half. He figured that such a short sentence was a reasonable price to pay for spending the rest of his life without money worries. He was tired of working so hard. Prison sounded like a vacation. One evening he simply told his wife and three children, "Guess what. I'm going to jail."

After the judge sentenced Senn to ten years, his lawyer promised him quick parole. The court let him spend his last free Christmas in Carmel, California. When the holiday was over, he was picked up by a Department of Corrections pilot who had been his first flight instructor.

"He let me fly myself up to Folsom. Then they transferred me over to San Quentin, where I was the warden's assistant. It was great up there in Marin County. The air's so clean, and I got a pass anytime I wanted to go visit the library. Conjugal visits were no problem. I had the run of the kitchen. The other inmates loved me when I persuaded the prison to install a computer terminal for instructional purposes. They couldn't believe it when I got the Veterans Administration to pick up the cost of data-processing courses on the G.I. Bill."

At every parole hearing, Senn believed he would ultimately be freed. But when he refused to give back the money, authorities refused to show clemency. Then he found out his lawyer had been working both sides of the street. A close friend of the company president, the lawyer passed along all details of Senn's case.

"What was particularly embarrassing," Senn recalls, "was the fact that I had given the attorney fifty thousand dollars to put out a no-hit contract on me with his underworld contacts. I had participated in some illegal dealings unrelated to my crime with the other officers of the company. I was afraid they might try to wipe me out to keep that part of the story quiet."

Senn believes this may explain why the company never pressed charges when he was paroled after serving five years of his sentence. "Sure, some of the banks I'd defrauded wanted their money back," he says. "And the IRS came around. But I was able to settle cheap. Those creditors who refused to take a nominal amount are just out of luck. They can follow me around to watch my activities, but that sort of thing gets to be expensive. As long as I liquidate slowly, there shouldn't be any big trouble. The whole thing did break up my marriage. It was hard to pick up when I came home after so long."

In the dining room the women he lives with and her three children are cleaning up after a late dinner. One by one, members of his new family check in to say goodnight. "It didn't go exactly the way I'd planned," he says, "but at least I won't end up apprehended on a Rio beach like the Lavender Hill mob.

"Right now I know of three similar jobs that would net ten million dollars with considerably less effort. But I'll leave that to the professional cons. Were you aware that the U.S. Agriculture Department has Leavenworth inmates programming millions in Commodity Credit Corporation payments? By the time they get out, they'll be ready to go in a new career."

SILICON VALLEY SPIES

By Christopher Simpson

Austrian physicist and businessman, Rudolf Sacher is middle-aged, well educated, and worth over $1 million. The only thing that stands out about him is his heavy, black handlebar mustache. He is also a secret agent for East Germany, according to Werner Stiller, a former colonel in the East German security service who defected to the West in 1979.

Stiller's defection touched off a chain of arrests that has resulted in the conviction of 12 West German government officials and NATO personnel. Austria is a nonaligned country, and its attitude toward Soviet Bloc agents is somewhat less hostile than that of West Germany. Sacher and his business associate Karl Heinz Pfnudl, who was also named as a spy, remain free to do business and to move about at will.

Sacher denies the charges leveled against him. Stiller made up the story, he says, to gain credibility with West German intelligence and the CIA.

A growing number of facts, however, suggest that Sacher has been a very successful agent for East Germany, one whose specialty is securing technical information on microcomputers, integrated circuits, and other electronic wizardry from the United States. Sacher, it seems, has been financing Peter T. Gopal, who in turn allegedly has been stealing the most advanced electronics secrets from such top American companies as Intel, National Semiconductor, and Zilog.

Peter Gopal is on trial for the theft of trade secrets from Intel, a leading microcomputer manufacturer, for conspiracy, and for bribery of employees of various microcomputer-research firms. According to grand jury testimony, Gopal was running a sort of black market operation in stolen technical information. Computer tapes and equipment used in manufacturing microcomputers had been purchased and illegally sold to the highest bidder. At least some of it apparently has ended up in the hands of the East Germans and the Russians.

Gopal's luck ran out in 1978, when he allegedly offered Tom Dunlop, an executive at National Semiconductor, a suitcase full of the latest Intel equipment for $200,000. Dunlop turned him in.

The Gopal case is only one episode in a vast Soviet effort to beg, borrow, or steal U.S. microcomputer technology—a task on which the Russians have spent well over $100 million since 1971. The espionage campaign ranged from simply buying U.S. technical magazines to a bizarre plot to obtain information and financial leverage by buying up American banks through an Asian front man going under the pseudonym of Amos Dawe.

The KGB has always paid special attention to international trade; few major Soviet trade deals are settled without direct participation by the KGB's Industrial Security Directorate. The first deputy chairman of the Soviet Chamber of Commerce, for example, is Yevgeni P. Pitovranov, a KGB general. Pitovranov, who frequently represents the Chamber of Commerce at international trade fairs, was KGB chief of station in Peking during the early 1960s.

At least two other KGB agents have recently been active in the United States in the guise of trade representatives. Vasili I. Khlopyanov, who works out of the Soviet consulate in San Francisco, was expelled from Thailand in 1971; Vladimir Alexandrov, the commercial vice-consul in San Francisco, was expelled from Italy in 1970 for spying. Both have aggressively sought out contacts in the electronics industry.

In all too many instances the Russians can simply buy sensitive high-technology equipment on the open market. They have purchased radar, computers, and lasers from willing Western businessmen.

Some sensitive technology, however, has always been barred from export. Equipment and know-how for the manufacture of microcomputers and integrated circuits, for example, have always been controlled, even in the heyday of détente.

In such cases the Russians' most effective tactic has been to employ unscrupulous import-export traders, some of whom work for prominent American companies, to buy high-technology products, disguised as washing machines, air conditioners, or other large appliances, for shipment to Eastern Bloc countries.

Few such white-collar smugglers have been apprehended, even though the Commerce Department, the Customs Service, and the FBI all agree that the practice is pervasive. One reason is that no one is concentrating on tracking them down.

Still fewer smugglers have been punished. In five cases of illegal electronics exports, originating in California, where the exporters were caught red-handed, not one prosecution resulted in a jail sentence. Fines have been slight, compared to the profits involved. One Canada-based buyer for the Russians recently was fined only $1,500 for the illegal export of electronic test instruments worth $1.5 million. Most of those companies caught with the goods are still in the electronics export business.

Russian electronics espionage is aimed primarily at the Santa Clara Valley, a cluster of suburbs between San Francisco and San Jose. Despite some competition from Japan and Texas, the valley remains headquarters to the most sophisticated electronics manufacturing and research operation in the world. More than 40 "Top Secret" projects are under development there, according to the FBI, and 400 more require "Secret" clearance.

Silicon Valley's chief products are the tiny chips used in computers—the hottest thing in electronics today. Controlling everything from automatic corn poppers to ICBM guidance systems, this new technology distills the power of a computer into a sliver of silicon the size of your fingernail. Twenty years ago a computer this powerful would have filled a room and would have cost millions of dollars. The cost of many standard chips is well under $20 and headed down.

The Russians—and the Japanese, the Germans, and just about everybody else—want to skip the billion-dollar investment in R&D it would cost to create their own chip-building methods. Instead, the Soviet Bloc has concentrated on reverse engineering or, when necessary, stealing U.S. chip-making technology.

Furthermore, Moscow has a particularly strong motive to keep an eye on Silicon Valley. Because the new microcomputers dramatically boost the capacity of American radar, missile-guidance systems, communications, spy satellites, and other military hardware, Silicon Valley is a vitally important source of intelligence on our military capabilities and on possible countermeasures to these weapons.

"About ninety percent of Soviet microcircuit production goes directly to the USSR's military effort," says strategic trade expert Dr. Miles Costick, who heads the Institute on Strategic Trade in Washington, D.C. He argues that the Russians are eight to ten years behind the West in microcircuit technology and that this is one of the few areas where the West still has a decisive military advantage. Obtaining these technologies, Costick says, ranks among the highest priorities of Soviet intelligence.

Rudolf Sacher has made a fortune in recent years by reverse-engineering U.S. electronics products, according to the Austrian business magazine *Profil*. His company, Sacher-Technik, specializes in peeling back the microscopic layers of a computer chip to see whether slight modifications could yield new products not protected by existing patents.

The practice is legal. In many cases, however, "reverse-engineered products are technically inferior to the originals," says Chuck McLeod, a U.S. Customs agent who specializes in electronics-smuggling cases. "The main market," he adds, "is the Eastern Bloc, which is prohibited from buying the originals direct."

Sacher is quite candid about his relations with the East Germans. "We have developed a method whereby we can place electronic circuits in one tenth the space required previously," he declares proudly. "East Germany gave us the order and financed it." Sacher also admitted in an Austrian court that in order to get that contract he had provided the East German intelligence service with a 74-page report on the state of the art in microelectronic technology—a practice he claims is common among companies that deal with the East.

Sacher denies obtaining secret information from Gopal, his man in Silicon Valley, and Gopal denies buying or selling microcomputer secrets. Gopal says he is being framed by National Semiconductor for the theft of computer tapes that National Semiconductor had stolen from Intel.

But the evidence—if it gets into court—would seem to indicate that Gopal is guilty. In a conversation that one National Semiconductor engineer, Larry Worth, secretly recorded, Gopal allegedly offered to sell him the Intel tapes and remarked that he had "just returned from Europe," where he sold similar stolen data. Gopal also offered to buy "anything you can get" from National and said that he had a fund of $1.8 million specifically for buying

American microcomputer equipment.

Gopal's business had just received a $1.8 million order for electronics parts from a Swiss company with close ties to Pfnudl, Sacher's partner. Shortly after Gopal's arrest, the Swiss company was liquidated.

The details of Gopal's foreign connections are unlikely to emerge at his trial, which has to do strictly with the issue of Gopal's alleged offer to sell trade secrets to National Semiconductor. Santa Clara County District Attorney Douglas Southard said "I think this case is clearly related to international espionage, but I don't have any need to spend my county's money prosecuting that." And the case has already become very expensive. Southard thought the FBI might be interested in the investigation, but the case had been under way for two years before he heard from the FBI. A spokesman for the FBI comments that his agency "lacks jurisdiction. It has never been an active case."

Conflicting jurisdictions and lack of support from the top consistently undermine attempts to control white-collar smuggling in high-technology goods, according to one Customs agent whose job takes him from one nondescript office to the next along the California coast. It's a problem faced by his colleague, Agent McLeod, who spent years tracing the activities of Gerald Starek and Carl Storey, two executives who eventually pleaded guilty to shipping circuit-manufacturing equipment to the Soviet Union through phony companies in Canada and West Germany.

"There is really nothing to stop them" from illegal exports, McLeod says. All it takes to set up an export company is "a telephone, some stationery, and a Dun and Bradstreet number."

The Customs Service is understaffed, he complains, and currency smuggling, drugs, and arms sales have higher priorities. "When was the last time you were searched leaving the country?" he asks.

The Commerce Department maintains an Office of Export Administration (OEA), which has primary responsibility for regulating high-technology exports. Although some Commerce Department investigators are considered the best in the field, the agency itself has a poor reputation.

The well-publicized sale of computers, automated welders, and other sophisticated equipment to the Soviet Kama River truck plant—one of the factories that provided trucks for the Red Army's invasion of Afghanistan—is only the latest in a long list of questionable sales to the USSR permitted by OEA. "With only eleven investigators working on the problem nationwide," notes one insider, "Commerce can't find a pay toilet in a men's room."

Despite its involvement in several recent, highly publicized antiespionage cases, the FBI does not seem to put a high priority on ending white-collar smuggling of electronics. At least judging from court records and from interviews with prosecutors, the FBI has not entered into the investigation of illegal-export cases originating in Silicon Valley. One agent who counsels electronics executives who have received threats relating to espionage activities was apparently unaware of several companies that maintain records of illegal exports to the USSR, even though those cases have been well publicized in the local press.

This may be changing, however. The FBI participated in the arrest of André Marc DeGuyter, a Belgian who made his fortune by selling classified computer codes to the Soviet Union's intelligence service.

DeGuyter was arrested last May at Kennedy Airport when he turned over a $500,000 check to a federal agent posing as a computer company executive in exchange for what were supposed to be computer tapes containing the "Adabas" source code, a sophisticated program for translating computer languages at high speed. Documents seized from DeGuyter at Kennedy included a $450,000 letter of credit from Techmashimport, a Soviet trading company. The FBI asserts that DeGuyter was also trying to buy the same Intel equipment that Gopal allegedly was seeking through his "consulting" business.

One of the most common electronics smuggler's tricks is "B Listing." The Commerce Department keeps two lists of items of equipment that are commonly exported. Equipment on the "B" list can be sold to the Soviet Union or other Eastern Bloc countries with few restrictions. Items on the "A" list, however, can be sold only with a special license, or in some instances not at all. The problem is that most items on the "A" list—special ovens for drying microscopic computer parts, for instance—have some kind of cousin on the "B" list—a regular industrial oven, in this example.

Electronics smugglers who know the ropes have an easy time switching the paperwork and loading the embargoed equipment on a plane bound for a nonaligned country such as Switzerland or Austria. From there the shipment, sometimes with another set of phony papers, is simply loaded on a jet bound for Moscow.

The case of Starek, Storey, and II Industries is McLeod's favorite example of how ineffective export control is. II Industries is a Sunnyvale, California, firm producing ovens, spinners, sorters, and other equipment used to mass-produce silicon chips. Sale of such equipment to the Soviet Union has always been banned.

Gerald Starek, president of II Industries, began cooperating with the Russians in 1974, when a Soviet agent named Richard Mueller promised him a $1.5 million sale. At the time II Industries was in a serious slump. What the Russians wanted, according to trial testimony, was a complete assembly line for the mass production of microcomputers. They were willing to pay a total of $35 million to obtain it.

Starek contacted John Marshall, a consultant who designs integrated-circuit and microcomputer assembly lines, and sent him to see Mueller in Germany. Marshall was being paid $5,000 down and $500 a day, plus expenses, for his services. Mueller sent Marshall to Moscow. "They were very secretive about what they were doing," Marshall commented later. "They wouldn't really tell me anything. It sounded like they were setting up a factory in Russia; in fact, a number of factories."

The CIA tipped off the Commerce Department about Mueller in 1975, and Commerce officials called II Industries to inform the company that they believed Mueller's order was headed straight for the

Soviet Union. II Industries officials replied that they didn't know anything about illegal shipping of electronics, but they promptly canceled Mueller's order.

OEA's policy of informing suspects that the government thinks they might be committing a crime is intended to help businesses that might inadvertently have broken a minor rule. It can happen easily. The Commerce Department's manual of export regulations is two inches thick. When a smuggler like Starek is being dealt with, however, the practice is like phoning a house that's being ransacked to let the thieves know that the police are on their way.

Instead of filling Mueller's original order, Starek set up two paper companies under new names in Montreal. The equipment Mueller had ordered was then shipped by plane to Montreal, from there to Zurich, then on to Techmashimport, in Moscow. The shipment was labeled as washing machines, industrial ovens, and air conditioners.

Even today no one knows just how much electronics equipment Starek smuggled out after the Commerce Department's warning. What is known, however, is that Starek and two other American electronics executives involved in the scheme— Storey and Robert C. Johnson—are each worth well over $1 million now, thanks at least in part to their sales to Soviet buyers. Although the U.S. government eventually caught Starek, Storey, and Johnson, their profits from smuggling were never seized. After three years of court battles the three executives and II Industries were fined $25,000 each with three years' probation.

"This is the standard," McLeod comments bitterly. "This is the worst that has happened to any of these guys. If you can make a million dollars and all you face is a twenty-five-thousand-dollar fine and probation, that's not a very bad deal, many people would say."

II Industries has been sold to Cutler Hammer Corporation, a manufacturer of top secret electronics equipment for the Defense Department. Starek is in business again as head of Silicon Valley Group, which also builds microcomputer-manufacturing equipment. The new effort is financially backed by Storey, Starek's lieutenant in the Russian scheme. Storey's export license has never been revoked.

Mueller, for his part, faces arrest if he returns to the United States, but he is rumored to be living in Europe. He, too, has not been barred from export or import trade with American companies.

The Russians are still trying to obtain equipment for their microcomputer factories. Not long ago agents in Boston seized yet another shipment of Moscow-bound wafer scrubbers and other microcomputer equipment, including some built by II Industries. This time it was being sent by way of London.

In a similar case, William Bell Hugle was caught in 1975 using false export declarations to ship a microcomputer assembly line to Poland through a bogus company that was supposedly going to manufacture digital watches in Singapore. Hugle is the head of an electronics export firm known as Hugle International. He was never prosecuted, Customs agents say, partly because they were unable to fund a trip to collect statements from Singapore officials.

Hugle International went bankrupt after the sale to Poland was blocked, the Commerce Department believes, and OEA dropped efforts to bar Hugle from the export-import business. However, only the California office was liquidated. Hugle International continues to do business from Tokyo. The company's Tokyo address is still listed on the Commerce Department's Foreign Traders Index, a computerized service meant to encourage international trade. Hugle is listed as an "agent handling semiconductors and related devices."

The Soviet effort to obtain a factory to mass-produce integrated circuits is much more serious, most experts agree, than simply smuggling a few sample chips out in the false bottom of a suitcase. "The Russians thoroughly understand the theory behind microcircuitry and have the laboratory ability to reproduce anything the West can build," one Pentagon expert who specializes in technology-transfer problems says. "But they lack production know-how and the technology to mass-produce it."

They might have caught up already if one incredible espionage plot had panned out. Aimed at Silicon Valley, the $70 million banking scheme was masterminded by the Moscow Narodny Bank (MNB), using a Hong Kong-based land speculator named Amos Dawe. Dawe, formerly a Chinese postal clerk, whose true name is Law Sheng Moh, built a financial empire worth tens of millions of dollars out of some lucky land investments and a mountain of credit from the Singapore branch of the MNB. That empire collapsed in the mid-Seventies when it became known that Dawe and several other prominent East Asian businessmen were in fact "cut-outs," or front men, for the MNB. That collapse left a string of lawsuits from Singapore to San Francisco as Dawe's business partners and the MNB fought over the wreckage. It is from those records and an interview with Dawe himself that the following account is drawn.

The Singapore branch of the Soviet state bank played a very aggressive role in Asian financial affairs during the early Seventies. By 1973 Dawe and his holding companies were already more than $60 million in debt to the MNB. According to Dawe, the MNB frequently promised him $70 million or more in new credit in a scheme to buy up a number of small U.S. banks. The scheme, Dawe was to testify later, was called the American Plan.

The CIA took the Moscow bank and Dawe quite seriously, according to documents obtained through the Freedom of Information Act. A hundred pages of heavily censored cables and CIA reports trace the path by which Dawe bound himself to the MNB.

Late in 1973 the MNB sent Dawe and its Singapore chief executive to look over available banks in the United States. The MNB needed Dawe, it seems, to take advantage of what was at the time a loophole in American banking regulations. Although the rules then required an examination of foreign companies buying American banks, no such probe was required of a foreign individual. Dawe could, to all intents and purposes, simply walk in, lay the money on the table, and buy a bank.

The MNB guaranteed the purchase and hid its interest in the deal by issuing letters of credit, which were laundered through two other banks. Dawe used the money to make down payments on the Peninsula National Bank, a small bank in the Santa Clara Valley, and two other California banks.

It is reasonable to assume that Dawe and his MNB advisers did not choose the Peninsula Bank only by coincidence: they paid almost twice the institution's book value in order to gain control of it. The MNB, acting through Dawe, by 1975 had paid off 80 percent of the bill for the Peninsula Bank and owned a controlling interest in the two others, according to the CIA. Negotiations were even under way for purchase of two more banks in the valley.

Dawe refuses to talk now about just why the Russians were so interested in those banks. "What if I were to tell you that this is done by both of them [the KGB and the CIA] all around the world?" he asks rhetorically. "What do you think is in it for them?"

The attorneys who later prosecuted Dawe for bank fraud are somewhat more forthcoming. "The Russians wanted easy access to U.S. dollars," says former U.S. Attorney John Lockie. "Dawe wanted the stability and reputation that come from owning an American bank."

But why the Peninsula Bank? And why pay almost twice its value when stronger banks were available at a lower price? For one thing, the bank would have given them detailed financial records of hundreds of American scientists and engineers and the companies they work for in the world's most concentrated center of secret electronics production. Such records, one industry source says, "provide a clear picture of people's vices and weaknesses, their family life, political views, and sometimes even medical history." Bank records usually paint an equally clear picture of high-technology companies and their contracts, product plans, and financial health.

Whatever the Russians' plans were, they were shattered in 1975 when a Hong Kong financial paper, *Target* (which is said to have close ties with the British intelligence service), blew Dawe's cover in a series of articles exposing his ties to the MNB. Shaken by the disclosures, the California State Banking Department finally began asking questions. The MNB cut off his credit, Dawe fled to Thailand, and his empire began to crumble around him.

Today Dawe keeps a low profile, using unlisted phone numbers and apartments rented under assumed names, as he plots his return to the world of international finance. Agents of the KGB, he says, beat him within an inch of his life in the hallway of a posh Thai hotel when he turned on his Moscow backers and sued them for withdrawing the credit he had been promised.

Dawe is apparently out of the espionage business, at least for the moment; it is clear that the Russians are not. The arrest of DeGuyter and seizures of microcircuit-manufacturing equipment in Boston are proof that the American microcomputer industry is still very important to Soviet agents.

"This is a rampant phenomenon right now," says Charles Letch, whose Advanced Computer Technologies Corporation was one target of DeGuyter's effort.

Peter Schnell, of the West German firm Software AG, says, "People working for the Eastern Bloc attempt to obtain this technology almost weekly."

Soviet progress in missile-guidance systems, in particular, warns of the impact that Soviet advances in microelectronics production would have. Despite the use of integrated circuits that are ten years out of date in the West, despite their equipment's added weight and lower reliability, Soviet scientists have developed systems that can place a MIRV warhead within 0.1 nautical mile of American Minutemen missile silos. Strategic trade analyst Miles Costick says, "The SS18 missile has emerged as the most lethal ICBM in the world." Mass production and application of microelectronics to military systems portend a qualitative advance in Soviet arms.

"We're going to see this stuff coming back at us," Lockie predicts. "They're not making automatic popcorn poppers with it, that's for sure."

CYBERNETIC WAR

By Jonathan V. Post

Welcome to World War III, the Cybernetic War, created by machines for machines. The soldiers in this conflict are, of course, computers, and by 1999 there will be roughly a *billion* of them either on or orbiting the planet. The arsenals of the Cybernetic War are stocked with cruise missiles, MIRVs, and smart bombs. Its battle plans utilize such tools as robotics, pattern recognition, coding and game theory, cryptography, and simulation. All depend upon one super-weapon, the indefatigable computer, which, paradoxically, is also the best hope for human liberation.

How did the computer get into this war? What is the war about? How will it end?

World War I, the Chemistry War, killed millions with the new chemical technologies of explosives, poison gases, synthetic fuels, and cheap steel. World War II, the Physics War, exploited the innovative physics of aerodynamics, radar, submarines, rockets, and nuclear fission. World War III, the Cybernetic War, is based on scientific advances equally known to the public and equally important to future historians, if any.

In Geneva, scientists recently announced the first containment of antimatter. Antimatter annihilates ordinary matter on contact, releasing more energy than the fusion processes in a hydrogen bomb. A beam of antiprotons, each at an energy of 2 billion electron volts, was bent into a circular storage ring, in a vacuum. Computers guided these clumps of antiprotons by altering the effects of powerful electromagnets. Quick electronic feedback kept the antimatter from touching the walls of the storage ring. The antiprotons circled at nearly the speed of light for 85 hours, then were directed at a target. Instant—and total—destruction. Computer technology, together with physics, is producing the most amazing tools for research—and yes, the most amazing military systems.

In 1979, we begin to see a pattern in the Cybernetic War. It is "Anything you can do, I can do better." The strategy of hardware escalation is a simple one. Whatever equipment *they* have, *we* must have at least the same.

America drop-tests the reusable manned space shuttle *Enterprise* from a Boeing 747; Russia drop-tests a delta-winged space shuttle from a Tupolev Tu-95 Bear bomber. How do we know? A Lockheed "Big Bird" spy satellite snaps photographs from 160 kilometers overhead.

Arab countries spend millions of petrodollars on

American airborne digital computers for avionics, navigation, and weaponry; Israel builds the Elbit System-80 for their own Kfir jets. Their former air force cheif of staff calls it "the best weapon-delivery system for single-seat fighters available today."

Government troops in major nations stockpile microcomputerized heat-seeking missiles, with which one soldier can down an enemy plane; terrorists begin to do the same. How do we know? Fragments of a Russian-manufactured infrared-detecting missile are found in the charred wreckage of an airliner on the plains of Zimbabwe.

In computer weaponry, whatever one side has, the other at least has on the drawing boards. Sometimes, as with the IBM 370 computer and the USSR's Ryad-2, they can barely be told apart.

Robots are not necessarily as cute as R2-D2 or C-3PO. Sometimes they are as deadly as U-235. Today's genuine robots are sleek subsonic assassins. They are better known by the name of cruise missiles.

Cruise missiles are refinements of the old-fashioned drones, or unmanned aircraft. Small, fast, light, and packing an atomic sting, they can be launched from land, air, or sea. Once in flight, they zip along at altitudes too low to be detectable by radar. Cruise missiles examine the landscape below and to the sides of their flight path, perform computerized terrain analysis, search for landmarks using the techniques of pattern recognition, and compare these results to their own internal maps. They plot their own courses and home in on their targets with dazzling accuracy. Thanks to the new science of robotics and artificial intelligence, cruise missiles strike within centimeters of their intended destinations, having searched with almost animal cunning from thousands of kilometers away.

Similarly, the MIRV, or multiple independent re-entry vehicle, is a nuclear-weapon-delivery system that depends on sophisticated computer guidance. When a MIRVed missile sweeps down from space, it tosses out a dozen separate warheads. Each warhead adjusts its own trajectory, performs evasive maneuvers, releases radar-confusing decoys, and plummets toward its own military bull's-eye. A "smart" bomb is one that uses sensors (such as television cameras) and a compact computer to mimic the human processes of perception and decision making, thus finding its target by planning, instead of blind luck.

But the MIRV has peaceful uses. Both the US and the USSR have launched scientific MIRVs at the planet Venus. Late in 1978, some 15 distinct payloads splattered at the shrouded goddess of love like shotgun blast. Of course, each piece of shot is a computerized scientific interplanetary probe. We dare not admit that World War III is an interplenetary war.

SIMULATION AND DISSIMULATION

Computers, unlike people, have no way of distinguishing between real and imaginary worlds. Their data may have come from sensors or may have been programmed by the most abstract calculations. A computer program may process information from actual measurements of solid objects or from artificial "microworlds" invented by creative human programmers. Rather than being a drawback, this is one of the greatest advantages of computer technology.

Through the computer, a person can dream up a universe with its own strange laws and structures. This begins to explain the almost addictive attachment many programmers develop for their machines and the reason that simulation has become a major part of military tactical training. A pilot can crash time and time again, learning from each mistake, if his computer simulator is an airplane that never leaves the ground.

There is nothing vague or imprecise in the computer dreams of simulation. Mikhail Botvinnik, chess grand master who held the world championship for thirteen years, explains why computers have an advantage in combat. "Man is limited, man gets tired, man's program changes very slowly. Computer not tired, has great memory, is very fast."

Northwestern University's CHESS 4.6 program on a Control Data 176 can beat the Russian KAISSA computer chess player. But who is winning the simulation race? In all likelihood, the United States will maintain several years' lead over all other nations unless Japan, also a contender in the robot race, makes its move.

There are no black-and-white squares inside the chess-playing computer, no kings, no pawns. Similarly, inside a Navy computer there are no miniature ships; inside an Army computer there are no tiny tanks. In each case, the computer does store and manipulate some information in its memory. The information is referred to as a *model,* although it is a model only in the abstract sense.

The model in a chess-playing machine might include a list of the positions of the pieces in a game situation, plus a mathematical description and evaluation of the legal moves available to each player. The model in a Navy computer might be another coded list of numbers.

These numbers could be the latitudes and longitudes of a fleet of destroyers or the schedule of flights from an aircraft carrier. Numbers might also describe the speed and maneuverability of the destroyers under varying conditions of weather and fuel consumption. The computer model for the aircraft carrier might list the communications frequencies of the individual planes, their altitudes, or their weapons status.

If the models are continuously updated to match actual ships and planes, they are enormously useful in battle situations. A jet fighter's computer can track many objects at once, at faster-than-human speed, and perform the IFFN function: identification, friend/foe/neutral. The computer application that automates the entire battle is known as C^3: communications, command, and control. If the ships and planes are imaginary, we are in a microworld where the computer calls the shots.

"Game theory," invented by Otto Morgenstern and John von Neumann, gave military strategy a solid mathematical foundation. Whenever two opposing players each select an option from a set of available options, and the outcome of each pair of decisions is known, a computer can recommend an optimum strategy. Such a strategy is usually a mix of options, with individual choices being made at random. Human lives may now depend upon the flip of a coin or the roll of dice.

The value of computer modeling is high, because it can help to predict the future. Simulation examines a set of alternatives and evaluates the possible outcomes of tactical situations. In chess, as in war, one ultimately makes one's move, and the effectiveness of one's plan depends on what the opponent does next. In chess, as in war, it is undesirable and impossible to test all tentative plans and legal moves. The chess master works with a mental model of the game, imagining what the opponent's countermoves and counterplans might be in a number of alternative futures. The commander works with a computer, subjecting imaginary helicopters to dummy runs, moving simulated troops through simulated battles.

Simulation cannot, of course, foresee the future exactly. A good simulation is still quite valuable if it can eliminate a few disastrous plans. Computer simulation has become an essential part of strategic and tactical planning in all major armed forces. Anything that tips the odds is studied. Trial-and-error is unacceptable when errors mean lives.

Simulation is never guaranteed. Japanese war games had predicted that America would surrender if Pearl Harbor was attacked. Pentagon computer simulations predicted that US incursions into Cambodia could cripple the Vietcong.

SPEED OF LIGHT

Electrons flow through computer circuitry at awesome speeds. The speed of computation is limited only by the speed of light, 298,117 kilometers per second. Compared to this, jets, bullets, and missiles are sluggish indeed. It seems natural for computer warriors to investigate weapons that strike at electromagnetic speed and that kill at the speed of light, too fast for any human intervention.

Lasers were developed late in the Cybernetic War. Lasers and computers have their futures linked in several ways. Lasers transmit data as fast as computers generate it. They already connect certain computers in high-speed digital communications links and will likely play an increasing role in military communications.

Lasers can be used as target designators. One soldier points a hand-held laser like a flashlight. Its focused coherent beam illuminates a particular tank or plane—any target selected by the soldier. A second soldier launches a portable missile, which homes in on the glowing bull's-eye. The microcomputer chip for laser-designated missile-guidance systems is much simpler and cheaper than the microcomputers in MIRVs and smart bombs.

Lasers can be used directly as weapons. High-energy lasers vaporize flesh or metal from many kilometers away. The Department of Defense has allocated over $200 million this year for high-energy-laser research. Also, the single largest source of funds for computer research today is DARPA, the Department of Defense's Advanced Research

Projects Agency. Universities are loath to question Pentagon largess.

American satellites have been zapped by Russian lasers since October 18, 1975, when one of our early-warning defense satellites was blinded by infrared radiation while orbiting over western Russia. The official explanation (a gas-main fire) is almost certainly a cover-up. The Pentagon is unwilling to admit the Soviet lead in hydrogen/fluorine chemical laser weaponry.

The Air Force has contracted with United Technologies for an airborne CO_2 gas-dynamic laser. Carried by KC-135 or similar aircraft, such a laser could destroy antiaircraft missiles in midair. The Navy has TRW building shore-based chemical-laser facilities for cruise-missile and fleet defense. The Army has opted for an AVCO electrical-discharge laser to be mounted on an amphibious assault vehicle. Hughes has contracted for computerized-beam aiming and tracking for the Air Force and Navy; Perkin-Elmer does the same for the Army. The Directorate of Defense Research and Engineering coordinates high-energy-laser weapon research with laser-fusion research and ballistic-missile defense.

Some of the world's largest computers are now used to study laser fusion. Elaborate simulations test the use of lasers to create temperatures and pressures now found only in the sun and other stars. It appears that computer-aimed lasers, connected to computerized phased-array radar systems, might be able to destroy incoming ballistic missiles. An LBMD (laser ballistic missile defense) debate similar to the ABM (anti-ballistic missile) debate of a few years ago seems inevitable if lasers and computers together shift the balance of terror toward defense. One Pentagon admiral with this in mind told the *Miami Herald* that Pratt & Whitney lasers would probably not be used on people: "Frankly, no, I don't think so. I think it would be the type of weapon you would use on a high-value target."

SECURITY AND INSECURITY

As with lasers, the early history of computers is clouded by security. In each case, lawsuits were filed by inventors who claimed to have built or designed the first of these revolutionary devices. Atanasoff has won a patent suit and claims to have built the first digital electronic computer. Gould has won a patent suit and claims to have sketched the first workable laser. Both claim that governmental secrecy and military security classification retarded communication with their colleagues. The same applied to nuclear-physics research in late World War II.

Cybernetics is a word invented by the multi-talented genius Norbert Wiener. Concerned with "control and communication in the animal and the machine," it is a mathematical theory developed in the early 1940s. Cybernetics describes the action of complex systems, whether in electrical equipment or in the human brain. The theory came just in time, for the electronic computer was being invented. Of course, the military motive was there from the start.

The theory began with the war-research group of Wiener, Weaver, and Bigelow. They were trying to build an automatic antiaircraft gun that would *not* fire a shell at where an airplane was. Rather, it would fire where the plane was predicted to go, outguessing evasive maneuvers. At the same time, in the USSR, A. N. Kolmogorov was solving the same problem, for the same purpose. Perhaps the Cybernetic War began here, in war-research laboratories during World War II.

Note this connection with ballistics, the study of bullets or missiles in flight. Vannevar Bush had already built the differential analyzer, a mechanical computer to solve some equations in ballistics, but it was too slow. German V-2 rockets were raining death on England, and the generals were wondering how much more effective rockets could be if some machine could guide them accurately to targets.

The race was on to build the first electronic digital computer. Aiken at Harvard, Goldstine at the University of Pennsylvania, and Von Neumann at the Institute for Advanced Study each led teams of engineers into the unknown. Turing had given England a head start. Due to security, nobody knew what Atanasoff had accomplished for the Naval Ordnance Labs at the Aberdeen Proving Ground.

After the smoke had cleared, the winner was acknowledged to be ENIAC, child of John Mauchly and Presper Eckert. The aerospace firm Northrop immediately contracted Mauchly and Eckert to design a special-purpose computer. This device would fit in the nose cone of an intercontinental ballistic missile. It would, in midflight, navigate by looking at the positions of the stars. A prototype, BINAC, was demonstrated in 1949.

General Groves and a vice-president of IBM seemed interested. The inventors, broke, joined Remington-Rand (later Sperry-Rand), developed the first commercially available computer, UNIVAC, and made their first major sale—to the Air Force.

GLOBAL SPIES

Computers have created a revolution in the ancient shadowy art of cryptography—codes and decoders. Anyone with a computer is now capable of communicating with anyone else with a computer by means of an absolutely unbreakable code. This was the dream of cryptographers in the employ of Alexander, Caesar, Napoleon, and Hitler. The techniques known as "trapdoor functions" and "Rivest coding" are easily explainable to any mathematician in the world. Using some familiar properties of large prime numbers, and using a computer to perform the calculations, two people who exchange a numerical "key" or password can thereafter intercommunicate with perfect security. The National Security Agency has tried to suppress this mathematics in civilian (i.e., private) quarters.

Security involves the controlled access and safety of persons, hardware, and information. Much of military information, government information, and business information is stored and processed by computer. Thus, security today is intimately tied to the computer.

James Angleton, former head of CIA counterintelligence, insists that US national security has been breached by the KGB. The Russian spies reportedly tapped into CIA archives in the central computer system at Langley, Virginia.

The General Accounting Office says that the Social Security Administration computers—which keep records on almost all Americans—are totally vulnerable to unauthorized snooping and tampering.

The NSA, ten times the size of the CIA, used giant computers to scan almost every telegraph, teletype, and Telex message sent through American borders. For several years, these computers automatically searched for keywords such as "missile," "China," and "assassinate." Messages containing keywords were recorded, and human operatives alerted. Illegal snooping on so vast a scale is impossible without computers.

The Cybernetic War emphasizes hardware and information, rather than soldiers and civilians. Superpowers routinely monitor one another's radar, microwave, and radio transmissions. Satellites eavesdrop, satellites snap clandestine telescopic photographs in ultraviolet and infrared, satellites with nuclear-powered radar probe the seas for submarines. This global flow of military data is coordinated by computer.

The public knows little of the Cybernetic War. New services present isolated facts, difficult to interpret out of context. Washington debates sales of AWACS to Middle Eastern countries. AWACS? Airborne warning and control systems, such as the Air Force E-3A or the Navy E-2C Hawkeye, are aircraft that carry computers and communications systems and serve as control centers in battles, executing and relaying orders. Carter prohibits sale of computers to the USSR. Why? The UNIVAC computer ordered by TASS is ten times larger than needed to manage the 1980 Olympics data and might be used for military purposes.

The public is told little about the Cybernetic War. The computer industry advertises the nonmilitary uses of its products. Universities teach computer science and computer business, but not computer war. Writers and critics, so articulate on the philosophy of artificial intelligence and on the unexpected home-computer revolution, are curiously tongue-tied on the major issues of war and peace. Technologists, on the defensive in conversations with anti-technological laymen, are reluctant to discuss military applications. In Russia, cyberneticist Shcharansky speaks out and is condemned to the Gulag Archipelago of prison camps. How many American experts, with less to lose, have as much to say, or show the same courage?

CYBERNETIC WAR: 1999

Let us venture some predictions. Assume that by 1999, the Cybernetic War has continued its hardware escalation without having degenerated into a thermonuclear catastrophe.

There will be roughly a billion computers in the world, almost all of which will be smaller than a large book. A third will be for business and science, another third will be in people's homes, and the remaining third will be in military weapons and equipment.

The typical soldier will be directed in the field by a computer. He will be supported by an airborne computerized robot. He will carry computer-designed, computer-manufactured, computer-aimed, and computer-actuated weapons. He will maintain secure jamproof communication through a surgically implanted link to a computerized network.

The major strategic weapons will be computer-directed beams of photons, nuclei, and antimatter. The major tactical weapons will be unmanned, as human reflexes are too slow for the battlefield control loop. Human judgment will still play a selective role in target-rich environments. Most tactical decisions will be made by machine, and most strategic decisions will be chosen by humans from alternatives presented by computers.

Ideological warfare will thrive. New combinations of satellite video broadcasting, subliminal data presentation, and computational psycholinguistics will blanket the globe in propaganda and counter-propaganda. Attitudes toward privacy will change, as details of a billion lives are stored within computer memories. People will be overwhelmed with information, and there will be a major struggle for access to knowledge, as opposed to mere data. Education will surpass entertainment in total cost.

Computer-designed materials will outperform natural substances in exotic applications. Computers will pilot millions of airborne vehicles, including ground-effect machines, lighter-than-air cargo vessels, and noncombustion rockets.

Paper will have been replaced almost entirely for news delivery, money, and commercial announcements. "Library" will no longer mean a building. People will read about the events of the Cybernetic

War, but not in newspapers or magazines.

The gap between the rich and the poor will grow, but the poor will be more aware of this and more capable of action. Resentment of computerized-police-containment actions will mount. Sabotage of computer-managed production and distribution systems will provoke increased robotic security. Many will see the conflict as Man versus Machine.

For the moment, it is machine against machine. The computer was born in a military context and has since permeated every niche of the military environment. There is nothing remarkable about the Cybernetic War in principle, except that it might at any time flare up and scorch this planet as thoroughly as a sun gone nova. A war in which hardware hunts and kills hardware can find human software caught in between.

And yet, paradoxically, the range of possible futures is expanded by the computer. If computers can advance the technology of peace as efficiently as they have advanced the technology of war, then people, and robots, will inherit the stars.

UNBREAKABLE CODE

By Roger Rapoport

Were you the sort of kid who loved to fiddle with a secret-code ring? Do you send messages that you wouldn't want business competitors to intercept? Perhaps you cringe at the thought of a tax audit. If so, you're going to love this.

For years now it's seemed that the Silicon Revolution would leave us all naked to the world. Anyone with enough nosiness, gall, and the price of a big computer can build an electronic data base that contains more information about us than we can remember ourselves. The insurance industry has done it. So have the credit bureaus. Some government agencies do little else.

Now the computers that helped rob us of our privacy are giving it back—with interest. Two cryptographic geniuses have made the breakthrough that code builders have dreamed of for centuries: They've invented a practical code that can't be broken. Once you've coded your information, no one—not the CIA, not the NSA, not even the IRS—can figure it out unless you've told them how. With the right programming, most home computers could code and decode messages. But without the key, even IBM's biggest number crunchers could work far into the next century without unscrambling them.

It's enough to make professional snoops weep. In fact, they've spoken out publicly against nongovernmental code research, interfered with patent applications, and even threatened university-based cryptographers with prosecution under the State Department's International Traffic in Arms regulation. Now the Defense Department is seeking the power to review articles on cryptography and to ban publication of any that it considers too informative.

This round in the battle between privacy freaks and code breakers got started when Martin Hellman, a thirty-three-year-old Stanford University professor of electrical engineering, linked up with another code junkie, Whitfield Diffie. Schooled in symbolic mathematical manipulations at MIT's Artificial Intelligence Laboratory, Diffie had left an industry job in California to search informally for the perfect code. After studying the classical literature, he camped his way across the country, visiting all the major centers of cryptographic research. Each night he examined the latest technical papers from university and corporate labs by firelight.

At IBM's Yorktown Heights, New York, lab, a scientist suggested that he look Hellman up back in California. "When I arrived in Palo Alto," Diffie recalls, "I called Hellman, and we each immediately found the other to be the most informed person in this field not governed by federal security regulations."

The problem they were trying to solve is lodged deep in modern code practices. Most coded messages these days are sent from one computer to another over telephone lines. For confirmation, they are also sent by courier. But that doesn't come cheap, and it often means delays when long distances are involved. A computer-wise thief who's wormed his way into a bank's message network can vanish with millions of dollars before anyone realizes that his orders to transfer the money weren't authorized. Worse yet for government cryptographers, there's always a chance that the courier will be intercepted or will defect with the message.

Then there are the electronic eavesdroppers. The National Security Agency has computers tied into long-distance telephone links all over the world. The moment a phrase suggesting a topic that interests the agency appears in a conversation, the NSA's tape recorders kick in. Similar equipment monitors data-processing lines here and abroad. Anytime someone makes a call or sends a wire, the NSA can listen in. New equipment will soon enable the agency to read mail, even before it's sent, by catching and interpreting an electric typewriter's vibrations with remote sensing equipment. And virtually anything the NSA can record, the agency's computers can decode.

Hellman and Diffie concluded that the major obstacle to secure transmission of data over teleprocessing networks lay in distributing the key, the instructions that tell the recipient how to decipher a message. "Traditionally," Hellman explains, "keys have been moved by couriers or registered mail. But in an age of instant communications it was unrealistic for computer manufacturers to expect customers to wait days for the code to arrive. What was needed was a system immediately accessible to users who may never have had prior contact with each other."

The idea of sending coded messages to total strangers seemed impractical at first. "In the past," Diffie says, "cryptography operated on a strongbox approach. The sender uses one key to lock up his message, and the recipient has a matching key that unlocks the meaning. As Hellman and I talked, we became intrigued by the idea of a system that used two different keys—one for enciphering and a second for deciphering. This method would operate like a twenty-four-hour bank teller. Any depositor can open the machine to put his money in, but only the bank has the combination to unlock the safe."

For a long time now messages have been translated into high-security codes by converting the words into numbers and then scrambling the digits mathematically. What dawned on Hellman and Diffie was that a class of extraordinarily difficult mathematical problems, known as one-way functions, acted like their bank machine. A practical code could be built on them. Users would be able to list their encoding keys in a directory so that anyone could send them a coded message. Yet only they would have the decoding key. Eavesdroppers would have no hope of ever decoding the transmission.

What made this practical was the work of Ralph Merkle, a young student at the University of California at Berkeley. Fascinated by the notion of a public-key system, he began working in one of his undergrad courses on a one-way function that could be applied to a code. Lying awake at night, he visualized a technique that would permit authorized users to decrypt messages that baffled eavesdroppers.

"The idea," he says, "was for A to send B a message in a million pieces. One of those pieces would be tagged so that B could use it to find the decoding key. But anyone else would have to sort at random through all the pieces to find the right one."

Merkle's approach did not impress his instructor, who considered public-key distribution "impractical." Unable to convince his Berkeley teacher of the system's promise, Merkle dropped his computer course. Then he wrote up his ideas for a computer journal. It rejected them as complete trash. "When I read the referees' criticisms," Merkle recalls, "I realized they didn't know what they were talking about."

In the summer of 1976 he finally found a sympathetic reception in the Stanford electrical department, and his work contributed to the breakthrough paper on the public-key system. Published that November, the article, called "New Directions in Cryptography," conceded that sending out a million pieces to foil spies searching for one that carried the key would be too expensive. Hellman and Diffie remedied this problem by letting each user place his encryption key in a public file, at the same time keeping the decoding procedure a secret.

Since then Ronald Rivest, an MIT computer-science professor, and his colleagues Adi Shamir and Len Adleman have made the code breaker's job even more difficult by using a new set of one-way functions. Their method builds encoding keys out of the product of two large prime numbers—numbers that can be divided only by themselves and by 1. This generates a figure hundreds of digits long.

In order to find the decoding key, it is necessary to "factor" this giant figure—break it down into the original numbers. It can't be done. Not even the largest computers can factor the product of two numbers with more than 50 digits. Only the recipient who knows the prime numbers used to build his encoding key can retrieve the message.

The public-key system also solves the other problem in sending coded messages: How do you know the signal does not come from an impostor? The Stanford and MIT teams have both produced a forgery-proof digital signature.

The encoding and decoding keys, though complex, are really just mathematical instructions that reverse each other. If the code were built on a simple arithmetic problem instead of on a one-way function, they might say something like "multiply by five" or "divide by five." The procedure can be used in either direction.

So to sign a coded message, you just reverse the

process. Encode your name with the secret key you ordinarily use to decode messages. The recipient then looks up your public encoding key in the directory and uses it to *de*code the signature. Since no one but you could have used the secret key, the recipient can be sure it was you who sent the message. And since the keys are based on a one-way function, the recipient still can't find your secret key.

This makes it possible to sign contracts over a computer network. If the sender tries to renege on the deal, the recipient need only produce a copy of the digital signature to back up his claim in court.

When the first public-key ciphers were announced, they dropped like bombs into the middle of a running battle. Six years ago the National Bureau of Standards decided to help out the banks, insurance companies, and others that were desperate for a way to keep their proprietary information secret. The NBS invited computer experts to develop a "data encryption standard [DES] algorithm" for computers. (An algorithm is the set of instructions by which you use the key to turn plain text into code and then decode it again.) And they invited the spooks from the NSA to evaluate the ideas.

The NSA, of course, couldn't be expected to have much interest in codes that it could not break, and a good many critics complained that letting the NSA work on the DES was like putting the fox on sentry duty around the hen house.

Their uneasiness grew when the NSA persuaded IBM, which developed the winning algorithm, to withhold the working papers used to develop it. The

NSA insisted that this was only a security precaution in the best interests of all users, but it looked to many as if the government was simply trying to lock up the algorithm's mathematical roots.

When computer scientists tried to publish papers suggesting that the new DES was breakable, the NSA tried to classify their work. One of the agency's employees, a man who once proposed to keep tabs on the 20 million Americans with criminal records by wiring them with transponders, even attacked the critics' patriotism in an engineering journal. The NSA finally agreed to meet with dissenters, then promptly destroyed all tapes of the confrontation. Inventors working on cryptographic devices found their patent applications classified and were threatened with prosecution for even discussing the equipment.

The NSA claimed it would take 91 years of computer work to break the DES key. According to Stanford's Hellman, however, "DES could be broken by an enemy willing to spend twenty million dollars on a computer that could test all the possible keys in less than a day." The DES key is a string of 0's and 1's, known as bits. It is 56 bits long. All you'd have to do to make it unbreakable would be to switch to a key with 128 or more bits. Since it wouldn't make the DES device much more expensive, why was the government being so stubborn?

"It occurred to us," Hellman says, "that the NSA wanted an algorithm that it could crack. That would prevent anyone else in the country from using a foolproof code."

With that controversy to prepare their way, the

public-key codes have received a warm welcome from just about everyone but the government. Some New York banks have already decided to reject the NSA-backed 56-bit encryption standard. An officer at Banker's Trust Company said his company refused to go along with the federal plan because it "did not meet all the bank's requirements." Bell Telephone has also rejected DES on security grounds.

These corporations may be better served by private companies now hoping to market coding devices based on the systems MIT and Stanford inventors are trying to patent. "Since we would share some of the royalties," Hellmann says, "some government people suggest our opposition to DES is motivated by self-interest. Sure, we would benefit if public-key systems go into widespread use. But the facts are that our method provides real protection and DES can be broken."

Rivest is already consulting for companies that hope to market foolproof systems. "What we want," he says, "is to develop an add-on encoding device for computer terminals that any user could afford. We're building a prototype now and working to see that it ends up in the marketplace." Bell Northern Labs, a subsidiary of the Canadian phone company, has hired Diffie to help make electronic eavesdropping more difficult. At the company's Palo Alto research facility, he is leading a cryptographic research group that wants to show callers how they can mask their identity.

Some computer experts, such as George Feeney, who invented the concept of EDP time sharing and who heads Dun and Bradstreet's advanced-technology group, voice concern about the practicality of these promised systems. "The unbreakable code is a brilliant piece of conceptual work," Feeney says. "These inventors have done an incredible job. But some of us wonder whether the process may turn out to be beyond the current state of the computer art. We still don't know how long it's going to take to get this dream going and whether the cost will be realistic."

The NSA, though, has already begun to whine about the prospects of companies and private individuals communicating over foolproof lines. The agency's director, Vice Admiral Bobbie Ray Inman, is so anxious that he recently broke official policy to go on record about this sensitive matter.

"There is a very real and critical danger that unrestrained public discussion of cryptologic matters will seriously damage the ability of this government to conduct signals intelligence and protect national security information from hostile exploitation," he complained. "The very real concerns we at NSA have about the impact of nongovernmental cryptologic activity cannot and should not be ignored. Ultimately these concerns are of vital interest to every citizen of the United States, since they bear vitally on our national defense and the successful conduct of our foreign policy."

Another NSA employee, Joseph A. Meyer, has warned his colleagues in the Institute of Electrical and Electronic Engineers that their work on public-key cryptography and data encryption might violate the International Traffic in Arms regulation. This law, which the government uses to control the export of weaponry and computer equipment, can even be invoked to thwart basic code research.

As a result, people like University of Wisconsin computer-science professor George DaVida, who recently tried to patent a new cryptographic device, have run into trouble. Although his work was sponsored by the federally funded National Science Foundation, the Commerce Department told DaVida that he could be arrested for writing about, or discussing, the principles of his invention. A similar secrecy order was issued to a Seattle team that had invested $33,000 to develop a coding device for CB and marine radios.

Protests from the scientific community persuaded the government to lift its secrecy orders in both these cases. At least for now, academics and inventors can continue to write and confer on cryptographic schemes. But the threat of renewed government harassment has complicated further research. Universities have agreed to defend professors against federal prosecution related to code research, but they can't protect students. As a result, some students have decided not to contribute papers to scientific conferences. In at least one instance Hellman had to shield two of his graduate students at Stanford by reading their reports for them at a meeting of the Institute of Electrical and Electronic Engineers.

It's too soon to know whether the government will move to block the use of the public key, but Hellman and his colleagues fear that young cryptographers may be scared away by Inman's tough admonitions. This could hold up the practical refinements necessary to make the unbreakable code widely available. A real chance to stop crime in the electronic society might be postponed indefinitely. With computerized theft increasing every year and computers controlling more of society's daily activities, this doesn't seem wise. But this issue appears secondary to Washington cryptographers, who sound as if they would like to reserve the public key for their own use.

"I'm not suggesting government agents want to listen in at will," Diffie says, "but I'm sure they don't want to be shut out. For them the perfect code is the one only they can break."

PART EIGHT:
AUTOMATED ART

CYBERFORMS

By David Lyttleton-Smith

There's a painting at the Frick Collection I never tire of,'' says New York writer and art thinker Bill Chamberlain. "Suppose I could literally enter the painting. Say that I could actually walk onto the canvas and move around in an environment given to me by Bellini or Giotto or Tintoretto. That would be an extraordinary thing to do. It's the highest example of technology used in the service of art.''

Chamberlain is one of a new breed of artists who anticipate a future in which art and technology become so totally merged as to be indistinguishable. A world in which the tools of the technician—wires, computer terminals, TV screens, and lasers—are molded by the artist's hands to express the same creative urges as paint and stone.

"Technology represents a new natural element, added to earth, air, fire, and water,'' says Otto Peine, director of MIT's Center for Advanced

Visual Studies, in Cambridge, Massachusetts. "It has become a natural part of everyone's life. So, when art deals with life, it must deal with, and through, technology. This is an essential ingredient of the present and the future."

New art, born of technology, says Gyorgy Kepes (pronounced keh-pesh), director emeritus of the MIT Center, instead of being cold and technical, represents a warm, contemporary humanism. "Art implies to be human in the most holistic and complimentary sense," he says. "And, in order to be human, one has to be aware of the essential problems of a certain epoch in history." In our time, he thinks, the central problems are technological, and the new art can describe them better than traditional methods can.

Some in the field, including Chamberlain, predict that computers will be the key technological development in the new art because they can free the artist from much of his production drudgery and allow him to concentrate on creative concepts.

"If we talk about art up to now, and back to Egyptian or Sumerian art," Chamberlain says, "the issue of having your tools—be they parchment and pen, chisel and hammer, or brush and paint— do even a portion of the conceptual work for you was never even there. That's a brand-new issue."

Others believe that the computer's influence on fine art will be eclipsed by the brilliant potential of laser technology. "The laser will utterly restructure our perceptual environment," Kepes predicts, "and transform the city from a place where we store human beings into an exciting dialogue between man and the world around him."

Whichever form dominates, technology-based art will flourish when technical tools are mixed into the raw reservoirs of creative inspiration by a new, but familiar, type of artist.

"I have a good friend," Chamberlain says, "who can play a terminal the way Vladimir Ashkenazy can play a Steinway. This guy sits down at a computer, and suddenly things begin to happen. He starts bopping around from system to system, effortlessly creating art forms that would take me months to do. He has a relationship with that computer. He's relating to it in a sublime way."

Possibilities arising from this sublime relationship include complex geometric and free-form graphics, which computers can generate and display in any medium. Artists like John Mott-Smith and Charles Csuri (pronounced chew-ree) have already utilized computers in this area because they can work at levels of speed and intricacy that the human hand could never attain. This raises the possibility for a unique version of time travel through computer simulation.

Beyond microelectronics, "a new technology of light" will convert urban skylines into orchestrated laser symphonies during evening hours, according to Kepes,. "We cannot shape the cities, make them something fancy, rich, and embracing, by using chisel and brushes," he claims. "We must utilize the optimum range of contemporary technology."

Already Barron Krody, laser designer at the Art Institute of Cincinnati, and such artists as Rockne Krebs have used lasers to transform public spaces into dazzling light environments. Krebs once created a phantasmagoria of mirrors and lasers spanning the Washington Monument mall. On a lesser scale, Krody has crafted light environments in limited spaces that draw the viewer into sparkling infinity. Creating space and depth through light makes even the narrowest city alley a place of delight and openness, Krody believes.

Our first view of atmospheric light transformation may come during the flight of the space shuttle. One of the vessel's projects will be the creation of a suborbital light show. A human-forged Aurora Borealis will shimmer across the sky in an outburst of light and motion that will be visible to everyone in the Northern Hemisphere. Further ahead we may see permanent orbiting art forms that produce bands of radiance in the night sky. Other, ocean-borne laser displays will unite sea and sky into immense dancing-light patterns.

Ultimately computers may control vast outdoor urban laser networks, just as they now regulate energy and other service systems. The computers could vary the rhythms, colors, and intensities of the displays, soothing or stimulating the urban complex, as desired by the artist.

Lasers also present the opportunity for bringing true three-dimensionality to art for the first time, through holography. Limited 3-D photos have been possible for years, but holography is maturing as an art form. Artists like Harriet Casdin-Silver and Rudy Burkhout create startlingly dramatic and highly resolved images that float in space. More recently the Russians succeeded in producing a rudimentary 3-D movie more than 90 minutes long.

The future holds promise for holographic and photographic systems that won't require light, capturing images with sound waves instead. These units will have the obvious advantages of working in total darkness as well as in broad daylight. Audioholography systems are currently in experimental use for medical diagnosis.

The greatest impact of technology-based art will be in the way we relate to art and our world. The new forms, Kepes says, will bring about "a reintegration of the artist into society." The inbred limitations of traditional museums and tightly framed circles of collectors will regress while art spans the globe, touching every life.

Peine tells a story that dramatizes the difference between the new art and the old: A well-known New York collector was scrutinizing an electronic light work. He paced around the piece for a while, looking somewhat puzzled, and then, making up his mind, exclaimed with exasperation. "What good is a piece of artwork if you can't buy it!"

"During the Sixties," Peine recalls, "there was a real attempt to domesticate the [technological arts] so that we could put them in a box, because of collectors and dealers. And we had light boxes and sound boxes. We had all these expansive art forms that we could put on the wall or on a pedestal. All this came out of the tradition of collecting objects."

The new art has since broken free from these bounds, and most people in the movement feel this has been good. "I've never liked the idea of visiting art in museums," says Howard Wise, owner of

Electronic Arts Intermix, a New York firm promoting new art forms. "That's the appeal of kinetic and electronic art for me. It's a part of every day life."

Within a couple of decades, according to Korean-born video pioneer Nam June Paik, video will be as common a personal art form as the 35-millimeter camera is today. Paik and Stephen Beck have independently developed the basic tool of video art: the video synthesizer. With this apparatus, an artist can draw on a television screen, creating any pattern of lines, forms, and tones that he wishes, without a camera. The synthesizer can also manipulate normal TV images.

An example of how the synthesizer generates artistic experience is Paike's tape of the poet Richard Ginzberg. As Ginzberg speaks, his synthesized image shimmers, radiates, and changes intensity in a way that expresses his mercurial, mystical nature better than words or simple pictures ever could.

The inevitable linkup between video synthesizers and satellites will make possible the instant worldwide broadcast of video art. In 1978, for instance, a video-arts display from the Documenta VI exhibit in Cassel, West Germany, was broadcast to more than 30 countries, including the Soviet Union, thereby reaching the largest audience ever to view a work of art simultaneously. Future developments might bring video to every corner of the globe.

On a more personal level, Paik feels that anyone will be able within a few years to record his entire life minute for minute and store it for later viewing. Just when one could play it back and how exciting it would be are moot questions, but the potential for historical documentation is obvious.

Even the nonvisual arts will be touched by the increasing interplay between technology and creativity. Bill Chamberlain's writing, for example, has been significantly altered by technology. The typewriter on Chamberlain's Greenwich Village desk is actually a computer printer, tied to a monitor screen and a memory. This system produces prose invented by RACTER, Chamberlain's program for synthesized computer literature. RACTER spins out limericks and suggestive short prose of a completely unpredictable—and frequently quite astounding—nature based on the vocabulary and grammatical rules that Chamberlain has given it.

The mastermind sits bemused in his chair while his machine spews forth creative writing. RACTER may not be ready for the Pulitzer Prize, but it demonstrates the new levels of interaction that technology and creativity has achieved.

Still history shows that today's conditions are not unique: technology has always exerted a substantial impact on the arts, consistently providing artists with new materials and better tools. Gutenberg's invention of movable type transformed the book from a rarity among the rich to a commonplace among the masses. Edison's phonograph made musical performances timeless. Film and video have brought actors before more people in a single night than they could reach in a lifetime of live appearances.

For the visual arts, technology has produced long-lasting materials, such as acrylic and polymer paints, nylon brushes, and synthetic canvas. The inventor of photography eliminated painting's role as the medium of portraiture and historical chronicle. More recently video, which brought the Vietnam War into America's living rooms, has supplanted photos as our historical window and now photography is veering toward pure art. In every age new methods of art have supplanted the traditional roles of older ones, enlarging the scope of human expression.

Yet today's integration of art and technology hasn't been entirely smooth. Artists and technicians ultimately make for a peculiar match. Though the two have much in common, they approach their systems from utterly different viewpoints. The artist is obsessed with possibility, the technician with efficiency.

"Art and technology have a very close connection," Howard Wise affirms. "Great artists and great scientists share many of the same qualities: tenacity, courage, and imagination." However, in practice, "psychology differences between artists and technicians," as Krody calls them, "spring up when the artist attempts something beyond the accepted boundaries of a technology." The technician's natural response is: "It can't be done." Often this is true. Other times, the collaborative efforts of artist and technician bring about an advance in the state of the technology.

Holographer Casdin-Silver recalls such an incident with technicians at American Optical Company. I came through with this crazy concept that wasn't at all possible in their heads," she says. "I

380

wanted to use garbage light, the extra light they worked particularly hard to get rid of, to reinforce the strength of my three-dimensional images. This was the exact opposite of their procedures." But in time Castin-Silver and the technicians produced a way of using the garbage light as she wanted.

In another instance Casdin-Silver wanted to create a holographic image that would float in front of the photographic plate, instead of behind it, where it had always been before. The technicians scoffed. "But I played with the problem," she recalls, "and by playing with it and feeling my way, I learned how to do it. When I brought my image out front, they were totally amazed."

Rudi Stern, an artist who uses neon in totally new and different ways, has encountered similar problems with technicians. Neon formers traditionally make flat, representational signs. When Stern asked the technicians to work in three dimensions, they were reluctant, but eventually their expertise and his creative imagination united to produce striking pieces.

Acquiring sufficient technical background has proved difficult for some new-mode artists. The most effective means of instruction up until now has been informal apprenticeship. Krody, for one, worked with General Electric near Cincinnati to create his works. Other artists have varying degrees of technical skill, but few have the breadth of background in their craft that a master painter could acquire in his.

One exception is Earl Rieback, a nuclear engineer turned experimental light artist. Rieback has no art-school background, and this he considers a definite advantage. "Few top artists have studied art," he maintains. "The top people in the field, the innovators like Jasper Johns, seldom have studied art. It makes them work within conventional constructs. Even when they desire to push forward, they have too many preconceived ideas."

Instead of learning art theories, Rieback believes that an artist should concentrate on learning everything about the materials and processes he is manipulating. An extensive technical grounding, he feels, is something that must become standard if tech-based arts are to achieve their full potential.

"Every artist must know his technology. It's necessary for success," he states. "What I know allows me to determine immediately how to accomplish something. My knowledge leads me correctly." Rieback makes his pieces faster, more cheaply, and more durably than his less well versed colleagues. For instance, his *lumina*—huge spaces that enclose patterns of swirling, shifting light—are cooled by natural convection rather than by fans, and so there is less chance of a breakdown.

As technical challenges are met and the abilities of new-form artists deepen, the synthesis of art and technology will create a universally accessible art form. "Technological art consists of movement, and movement takes the place of subject matter that we once could identify with: cows, madonnas, milkmaids," Wise says. "When art became abstract, it became less accessible to the average person. Kinetic and electronic art can be understood by everyone."

Rosemary Jackson, directory of the Museum of Holography, in New York City, stresses the openness of technological art. "Technology is logical, not mysterious. If you can't understand what an artist is doing, he simply hasn't done a good enough job."

The use of forms everyone can understand—a populist art for the new age—firmly reintegrates art and society. The new art's accessibility will help make our complex universe comprehensible to us again. For example, many of the programs that generate graphics in computer art are based on natural phenomena, such as atomic-particle motions, sine waves, and pendulum motion. This use of what Jackson has termed "the pulse of life" may lead us on an aesthetic journey into the new universe that science has revealed.

Years ago Buckminster Fuller's discovery and promotion of the tetrahedron as the fundamental geometric structure of material phenomena supplied us with physical models to describe the fourth dimension. Now the work of technological artists will help us understand worlds still being discovered by physicists and may ultimately give us visual analogues of a universe previously known to us only through equations.

Working with a concept that Kepes calls transference, today's artists translate conceptual information from one sense form into another. Visual information, such as geometric curves, can become analogous audio experiences or thermal experiences, all three determined by a core of generating data, such as a number series.

Astronomers have already converted radio signals into visual simulations of the big bang. When all sensory data can be similarly switched, we will have entered a new era of knowledge through the marriage of technology and art.

Originally the tools of painting were used to decorate surfaces and perform other mundane tasks. No one would have imagined that one day artists like Rembrandt, Rubens, and Picasso would express their deepest creative longings with these artisan's tools. Perhaps today's techno-artists should be considered the forerunners of tomorrow's Michelangelos and Leonardos.

The final development in technological art, according to Krody, may be the total elimination of media in favor of brain-stimulation devices that generate music, colors, tastes, and tactile sensations within our minds. This will represent the true link between artist and audience—the transmission of pure creative thought.

Imagine how such an integration would affect us and our institutions:

It's 2025 and they've closed down the museums. The works of the masters lie in the darkness of subterranean storage vaults. People don't need to see them anymore. Why should they? In the privacy of their homes, they can view the entire collections of the Louvre on 3-D entertainment modules or visit Michelangelo's Florence via electronic brain stimulation. No more hassles with crowds of tourists or with pompous tour guides.

Great art no longer exists in only a few places; it has moved out of the academies and into the world, enriching both the intimacy of the home and the expanses of the sky. Cities on the ground—and in space—are now wrapped by bands of air space devoted to cultural events: massive displays of colored gases, flying sculpture, and suspended prisms, which bathe the environment in kaleidoscopic spectral rainbows.

Home computer systems craft 400-page novels in hours, suited to the owner's personal taste, or duplicate the *Mona Lisa* from recyclable materials so that only a chemist could distinguish the copy from the original. Dreams can be stored in the computer and be played back the next day for wide-awake 3-D perusal.

Is it any wonder museums have faded away?

SOUL OF A NEW MACHINE

By Robert Weil

Best-selling books are no longer just written. They are created by marketing experts. A young author's prose may suggest the literary talent of F. Scott Fitzgerald, and his manuscript may boast the commercial appeal of *Gone with the Wind,* but unless a financial consultant determines that the book can sell, it will become lost in the pile of 40,000 volumes published annually in the United States.

Occasionally, though, a book miraculously slips through this Calvinistic process of predestination, becoming a commercial and literary success through the craft of the writing and the narrative.

The Soul of a New Machine, published quietly in the summer of 1981 (Atlantic–Little, Brown; $13.95), is such a book. Its success, ignited by word of mouth, then fanned by the encomiums of reviewers, bears testament to the fact that fine writing can succeed on its own. It signifies also that science books have come of age.

"It's the kind of thing you dream about," remarks a dazed Tracy Kidder about his unexpected success. His story about the design and the construction of a 32-bit minicomputer by Data General in Massachusetts reads like a fast-paced thriller. His characters, the talented young engineers known as the "Hardy Boys" and the "Microkids," come across as masterminds in a James Bond novel. *The Soul of a New Machine* banishes the stereotype of the scientist as a staid, bespectacled type, tediously laboring for decades over his laboratory research.

Kidder's project began innocently enough in the fall of 1978. He had been compiling lists of possible book subjects, and he went off to see his editor at Atlantic–Little, Brown.

"Hey, why don't you look into computers?" the editor suggested, and then he gave Kidder the name of Data General's Tom West, the orchestrator and leader of the Eagle computer project. It was a strange assignment for the Harvard-educated English major, whose background was "absolutely nonscientific. I didn't even have a basic knowledge of computers," Kidder admits, but "I don't think that that's necessarily a disadvantage. I was not writing for a completely computer-oriented audience."

The Hardy Boys had already completed the design of the new computer when Kidder arrived at Data General, in Westborough, Massachusetts. "They were getting ready to start debugging," he remembers. "So I had to reconstruct the design phase."

West introduced him to all the lieutenants, and Kidder "began to hear their stories on how it all came about." He became friends with many of the people involved in the project and after several months of observation acquired an intimacy with the project and its architects that is unusual for a reporter covering a story. It is precisely this intimacy that distinguishes the book and enables lay readers to comprehend the powerful dynamics that characterize American technological research at its best.

Trouble arose after eight months of observation. Apparently the fact that Kidder had gained the confidence of those who were creating the computer began to worry the corporation's executives, who feared that trade secrets might be divulged. "Someone upstairs got cold feet, and I had a few go-rounds with the chief counsel," he recalls.

"Some people there wanted to get more control over me than I thought I could possibly surrender." Although the lawyers eventually resolved the situation, Kidder rarely returned to the Data General plant.

The discord that developed between Kidder and Data General has been forgotten in the wake of the book's appearance on national best-seller lists. "The book's success is not hurting their business," he adds dryly.

What emerges as a dominant theme throughout *The Soul of a New Machine* is a respect for the tradition of fine craftsmanship. It's a time-honored New England trait, personified not only by Kidder, who takes considerable pride in the craft of his writing, but also by the Microkids, who assembled their computer at an astonishing pace.

One is even reminded of the Massachusetts Bay Puritans in the seventeenth century. They zealously operated their community for the common good, believing that the misdeeds or slothfulness of *one* citizen would inevitably jeopardize the safety and survival of the fragile community that had been carefully constructed under the eye of God.

West may not have acquired his managerial brilliance through a study of Puritan philosophy, but his manipulation of the computer engineers compares well with the workmanship of an earlier time. "He [West] was very good," Kidder says enthusiastically. "He had this little competition set up. He talked about things like peer pressure. 'If I screw this up, I'll be the only one, and I'm not going to be the only one,' was the prevailing philosophy. There was a lot of this kind of manipulation."

Understandably, Kidder remains optimistic about the future of American technology. "We still have Bell Labs. We have the Watson Research Center. Those are pretty formidable operations. I think IBM is formidable as well. Think of the billions of dollars people lost underrating IBM. It's astonishing."

At the same time he is concerned about the direction of American industrial technology. He believes that the largest corporations are sacrificing long-term development in favor of short-term profits. "I really think that the way to ruin an industry is to make it boring, to start looking for the guaranteed return, particularly in research and development. It seems contradictory to even think that you have to have guaranteed returns with R and D."

While the book reads like a paean to the modern computer, Kidder is personally suspicious of the computer industry. It is a paradox that should not go ignored.

"Computers are used, I think, to a large degree by people in power to stay there. . . . The computer *revolution* is the wrong word, in many cases. Computers have been a prop of the status quo. The stock exchange is the perfect example," Kidder points out.

"The irony is," he continues, "that you can do terrific things with word processors, but unless you can cure the bureaucratic mentality that loves papers that no one ever gets to read, all that's going to do is just make the problem worse."

Kidder further questions the "unjustified mystique" of computer technology and is worried about the "terrible ways" in which the products of, say, California's Silicon Valley may be used. "There seems to be no controlling them."

He is also uneasy with the notion that the craftsmen who designed the Eagle computer are putting other folks out of work. "I suppose it's an old story—craftsmen exercising their craft at the expense of other ones," he remarks. "The obvious thing that comes to mind, of course, is the printers. Printing is a craft that computers have pretty much wiped out. . . . I also think of office equipment. There are a lot of good ways to use those wonderful word processors, but some ways of using them look like they're just going to take away whatever's interesting in a secretary's work.''

True to his philosophy, he has no desire to own any sort of home computer. ''I don't think I want to write anything on a word processor. . . . I'm set in my ways,'' he says.

The success, however, of the book—the national television shows, the front-page review in the *New York Times Book Review,* and the sale of the book to paperback for a figure rumored to be close to half a million—has left thirty-six-year-old Tracy Kidder unsettled in many of his ways. He even lost 15 pounds after doing one show for PBS television. Life for him in the hamlet of Williamsburg, Massachusetts, may not be quite as undisturbed as it once was.

What's in store for him? Kidder says he's not going to abandon science and technology, but he might try another subject. For now, he's going back to ''just making lists.'' ''I need a little rest,'' he muses.

SILICON ORCHESTRAS

By Spider Robinson

One of the nicest things about the silicon revolution is that miracles are coming into the price range of the ordinary consumer. The latest field to benefit from this boom in cheap computing power is music. Now for less money than you might pay for a good piano or a decent, second-hand car, you can have at your fingertips the equivalent of a whole orchestra in the form of a compact but powerful electronic synthesizer, a computer-controlled music maker of a thousand voices.

Synthesizers of course have been around for years. Today they are more sophisticated and ever more expensive. A few years ago a New York firm, Digital Keyboard, Inc. (a division of Music Technology, Inc., 105 Fifth Avenue, Garden City, NY 11040), came out with what is generally considered to be the Rolls-Royce of synthesizers, the General Development System, or GDS. Larry

Dunn, of Earth, Wind and Fire, plays one on the new Stanley Turrentine album, and Michael Urbaniak, Tangerine Dream, and several university music departments are quite happy with theirs. Fully digital, user-programmable, and equipped with more than 1,500 "voices," or distinct musical sounds, the GDS has at its heart a computer powerful enough to handle an orchestra's worth of playing and still manage a small financial empire on the side. And you'd need one to own a GDS. It currently lists for $30,000.

This year Digital Keyboards unveiled a new synthesizer that gives electronic musicians much of the power and the range of the GDS for a lot less money. This new model, Synergy, lists for one sixth the price of the GDS, or $5,000. Even at that price what you get is a very advanced, very sophisticated machine.

Synergy's computer is capable of producing uncanny duplications of acoustical instruments, down to the scratch of the bow or the pluck of the string. It is also equipped with a 74-key keyboard—remember Bach had only 61 keys to work with and Mozart, 73—which is, in the jargon of the trade, velocity and pressure sensitive. Very simply what this means is that a player can modulate the volume of the music as a pianist does when he adds texture to the music by either gently depressing or slamming down on the keyboard. Previously this was available on only the most expensive synthesizers, like the GDS.

In addition, a Synergy musician can play anywhere from 8 to 16 notes at the same time by using as many as four different voices that can be interchanged among the notes without getting the machine mixed up.

Beyond the standard 24 voices built into the machine, a Synergy player has the add-on option of a series of special 24-voice cartridges that fit into a slot similar to the ones used on video games for game cartridges. This gives Synergy the full range of the GDS's 1,500 voices, many of which have names as colorful as their sounds—names like Nasalfuz, Bellorgy, Tinkbell, and Pigpluck.

For an opinion on how this high-tech music machine works, I talked to Clark Spangler in Los Angeles. In the past, Spangler has done considerable design and development work for some of Digital Keyboard's biggest competitors, and it is no coincidence that Yamaha's CS series of synthesizers carry his initials. You no doubt have heard Spangler somewhere, either on the more than 50 albums on which he's played with people like Miles Davis, Stevie Wonder, the Band, or Dizzy Gillespie, to mention but a few, or in well over 200 movies and TV shows.

In Spangler's view, the Synergy is a promising development for the music-and-synthesizer field, gaining it more serious acceptance. "As high-quality high-tech machines come into the range of anyone who can afford a decent piano, more people will come to realize that the synethesizer is not an intimidating electronic device, but a true musical instrument," he says.

The biggest boon of machines like Synergy, Spangler adds, is to composers. With a sophisticated synthesizer at his command, a composer has instant access to an electronic orchestra and "to textures, timbres, and colors that Beethoven couldn't even fantasize about."

COMPUTER GRAPHICS

By Robert Rivlin

Use computers to assist in advanced automobile design? Of course. Computers-simulate images of outer space? Naturally. Computer-generate special effects for motion pictures? Just look at *Tron*. But children's Saturday morning cartoons such as *The Flintstones* and *The Smurfs* created with the assistance of an advanced computer-graphics program? The idea seems almost ridiculous, especially given the frivolous nature of the cartoon images themselves.

There is, however, nothing frivolous about the experimental project now under way at Hanna-Barbera Productions, in Hollywood, one of the country's biggest producers of cartoon animation. For the past year Marc Levoy, Chris Odgers, and Bruce Wallace, who had been working on the Cornell University computer graphics program, have been writing software for a computer system at H-B to help create animated sequences with Fred Flintstone and the rest, using the computer's database and numerical processing abilities. And though Levoy, Wallace, and Odgers consider themselves more mathematicians and computer programmers than cartoon animators, they are proving once again that computers can play a central role in the creative process.

The Hanna-Barbera project is under the personal supervision of company president and cartoon-industry giant Joe Hanna, who originally commissioned the experiment and still maintains some ties with the Cornell University program. But the ties are those of shared information, not physical setup. Integrated within Hanna-Barbera's extensive production facility, the computer-graphics project bears little resemblance to the sterile, clean-room environment in which many computers are found. Space was set aside within the existing studio facilities, and the computer project moved in.

Indeed this reflects the computer project's aim. "We're not out to replace the cartoon animator," Levoy emphasizes. "Our goal is just to relieve him of some of the horribly tedious, boring, and repetitive work that normally goes into hand-coloring each of the cels in an animation sequence." The computer-graphics process also permits the direct recording of cartoon sequences onto videotape instead of having each frame first shot on motion picture film.

The first step in the experiment was an intensive study of the conventional animation process—a technique that has changed little since the first films were made. The illusion of movement, of course, is created by changing the drawing slightly from frame to frame; the eye's persistence of vision retains the previous image when the new one is presented, and the visual cortex perceives the difference between the two as movement.

One of the first programs written by Levoy, Odgers, and Wallace at H-B was an "electronic exposure sheet"—a word-processorlike system to help keep track of the thousands of individual pieces of artwork that go into a final production. (Each second of finished film has 24 frames, each requiring drawings for both the foreground and the background, which are created and manipulated separately.)

The exposure-sheet system also helps keep track of the different parts of the cartoon characters themselves; in Hanna-Barbera's limited animation process, only the features of the character that are actually moving change from frame to frame. Levoy shows us a set of cels representing Fred Flintstone talking (the word *cel* comes from the cellulose out of which the acetate drawing sheets are made). Sandwiched together as they would be on the animation-photography stand, Fred's components look like a complete figure. But he is actually made up of several different cels, some representing his facial features, some his arms and legs, some his torso. To make him talk, the animator doesn't have to change any of the cels except those representing the face. Then when Fred is done talking, a different set of changes will make his arms or his legs move. By not having to redraw the figure completely for each frame, hundreds of hours in this labor-intensive industry are saved each time a cartoon is produced. An even greater saving is accomplished by using the same set of cels whenever the character walks in a particular direction, shooting the same cycle of cels—the series of eight or nine leg positions in a single stride—over and over again as the character continues to walk.

Cost consciousness is on everyone's mind at Hanna-Barbera, and it may be one of the most compelling reasons why the computer might become fully integrated into the operation. Levoy estimates that a Walt Disney cartoon such as *Pinocchio* or *Snow White,* which was made in 1930s for $2.6 million, would cost around $15 million to produce now if the same level of animation were used. In those two masterpieces every frame of every character was drawn individually, with backgrounds sometimes consisting of as many as five separate planes, each moving separately to give an astonishing simulation of depth. Network television, the main purchaser of cartoon animation today, will not spend the money to support this kind of animation, and Levoy and the others in the computer-graphics project know it.

Sitting down to demonstrate the computer's capabilities, Levoy first enters the animation artist's outline drawings of the character's actions into the computer. "At this point," he explains, "the drawings are part of the computer's database—like any other set of shapes stored in a grid coordinate pattern."

As if they had been created digitally in the first

place, the drawings can be manipulated and changed. And, most important, they can be electronically colored by using digital "paint" programs that allow the artist to treat them as shapes in an electronic coloring book. In the conventional animation process, this is one of the most laborious tasks, requiring an artist to hand-color every element of every cel. In H-B's electronic process, however, the artist uses a penlike electronic stylus to pick a color from an electronic array presented on a TV monitor. He then simply touches the stylus with the area he wants colored, and the computer does the rest, filling the area with the selected shade until the boundary lines are reached. To choose another color, the artist simply returns to the color display and touches the stylus to another of its "paintpots."

"It's exactly like real painting," Odgers observes, "except that there are no drips or smears or differences in paint intensity from one cel to the next. And you don't have to clean your brush."

Underlying this apparent simplicity—and completely invisible to the artist—is a special type of computer memory known as a framestore or frame buffer. The picture area on the artist's TV display monitor reflects the organization of the framestore into at least 262,144 individual memory locations called pixels (picture elements). Each pixel can be displayed in any of the millions of colors of which an electronic color scheme is capable (infinitely variable degrees of red, yellow, and blue). The artist, using the electronic stylus, can touch any set of pixels, turning it any color he wants.

The next piece of computer magic Levoy performs is the electronic reassembly of the character's various body parts. Again replacing one of conventional animation's most tedious tasks—the physical photography of the animation cels after first sandwiching the appropriate pieces together on an animation stand—the program is based on some highly complex software written by Wallace, the rival of any work being done today in other advanced computer-graphics operations.

Using the electronic equivalent of a conventional animation positioning grid, the animator assigns each of the body pieces a coordinate in the frame, together with a set of instructions that tell each piece how it fits together with the other pieces. The eyes are defined as completely opaque, for instance, and fit inside the shape defined as "head." The computer now takes the digital information representing the various elements and merges the pieces, performing special operations along the boundaries where two shapes meet so that the "seam" is completely invisible. The electronic operation takes but a fraction of a second.

In a final, almost instantaneous step the character is merged with background art that has also been scanned into the computer, using the same kinds of computer programs. Again the computer treats the complex background image as it does any other part of its database; the strings of digital information representing the pixels could as easily be data representing checkbook balances being combined with various income statements.

The completed frame with character and background is then put on videotape. Next, in accordance with a preprogrammed sequence of instructions entered in its memory, the computer slightly changes the background, selects another computer-colored cel, and sends another frame to the video recorder. And the figure becomes animated.

To show the versatility of the system, Levoy asks the computer to generate a cartoon from pieces already in its memory. A little later the playback tape shows Fred Flintstone and Sylvester the Cat being pursued through the interior of a building by a whole flock of Tweety Birds, the image betraying no sign that it was created wholly within the computer's database.

"It seems farfetched right now," Levoy concludes, "but we think the day is going to come when Hanna-Barbera's shelves will be lined not with film cans and stacks of cels, but with computer disk packs and videotape instead."

Levoy, Odgers, and Wallace are not alone in their thinking that computer-created animation might revolutionize cartooning for television. Nor in fact is the H-B experiment the ultimate to which computer animation might be pushed. Even more highly advanced work is being done at the New York Institute of Technology, where programs were originally written by Alvy Ray Smith and Ed Catmull. The two have since moved on to head a special research group at George Lucas's Lucasfilm facility to see how computer-created effects might one day be integrated into Lucas films, such as future Star Wars episodes.

Smith and Catmull's work involved two fundamental animation techniques not tackled at Hanna-Barbera: electronic inbetweening and electronic background painting. The latter is somewhat similar to H-B's cel-coloring programs, except that it offers the artist a much wider range of creative choices. The stylus can be transformed into an electronic brush of virtually any thickness, or into an electronic pencil, crayon, or even airbrush. Pieces of the electronic image can be "cut out" and moved around, rotated, compressed, repeated, or enlarged —by touching the stylus to a "menu" of graphics functions displayed alongside the image.

This type of digital art system, originally conceived as a means of electronically painting background images for cartoons in the same way that the characters themselves were created, has found its way into commercial television. CBS is among those that regularly use a digital art system for the preparation of graphics for nightly newscasts.

Electronic in-betweening makes use of another of the computer's abilities—the calculation of a smooth-line transition between two sets of coordinates. In the conventional animation process, the highly paid animator actually makes only a few key drawings for each sequence—the images representing the beginning and end points of the movement. Lesser-paid "in-betweeners" are then given the task of creating the succession of drawings that move the character between the key frame positions. The electronic program, however, can eliminate the inbetweeners altogether. Only the key frames are scanned into the database; mathematical-interpolation programs then move the character from one key frame position to the next in whatever time the animator specifies.

Where might all these developments take us? Television, with all its budgetary limitations, may have reached its ultimate computer-graphics level if the H-B project becomes a production reality and is put on line to help create some of the tens of thousands of feet of animation footage H-B puts out annually.

But not so the movies, in which there seem to be few budgetary restraints. Digitally created special effects have already been seen in Michael Crichton's *Looker* and abound throughout the Walt Disney Productions film *Tron*. But the next evolution will come soon—most likely first from the team of Smith and Catmull at Lucasfilm, but not quite in time for the *Revenge of the Jedi* episode of Star Wars.

Imagine that the background being scanned into the computer is not a silly cartoon image but a rendering of an alien planet's snow-capped mountains. By programming the computer, the artist is able to shift the light intensities slightly every time the image is displayed, creating an eerie, flickering effect in the methane snow. Now imagine that instead of Fred Flintstone or Barney or Sylvester the Cat, the foreground image being scanned into the computer is footage of the actors and actresses. The same programs that enable the merging of the cartoon characters can be applied to the real scene, placing the performers into a digitally manipulated image of the alien planet.

Finally the datastream of the composite image is

used to modulate a laser beam that translates the data into a 35mm motion-picture frame, the system's resolution so high that it is above the resolving power of the grains of film themselves.

If the television and film media are in the business of creating illusions and magic, then surely the best is yet to come.

PART NINE:
A SKEPTICAL EYE

LAST WORD

By Richard Ballad

Dr. Robert Jastrow, NASA: It is entirely possible that man has evolved as far as he can and that a new form of life, based on silicon, will replace the carbon-based human life form as the dominant species.
Stunned reporter: Do you mean a computer society could supplant mankind as the highest form of life on Earth?
Dr. Jastrow: Yes.

This snippet of conversation between me and Dr. Robert Jastrow two years ago has been sticking in my throat like a hair ball. It's been further irritated by epic battles with certain computers belonging to the telephone company, with various credit card machines, and with the videotape-editing devices at NBC News, which can do in ten seconds what it once took humans half an hour to sweat out.

But for all the cute and clever things these silicon-based computers can do, I find it appalling that we are beginning to lean on these weird creatures for decisions that should necessarily involve the human will and imagination. It's insanity to let these dumb machines decide who is a good loan risk, who will make the best employee, and whom we should date or marry. I don't care what Dr. Jastrow says.

Consider the world of gambling. The big casinos use computerized odds. They know they are going to make 14 percent, or 19 percent, or whatever on the blackjack tables and slot machines and maybe 90 percent on roulette. There's no romance in that. Compare the dullness of a casino manager's life with the gizzard-twisting feeling you get when you put your quarter or, better still, a slug into a slot machine and hit the jackpot because something inside told you this was the right moment and the right machine.

Computers, much like loan sharks and medical doctors and such trash, deal with projections based on probabilities. But it's the *possibilities* that make us want to get out of bed in the morning and try it one more time.

Imagine what history might have been if our ancestors had paid homage to a bunch of silicon chips and depended on computers to tell them what to do. Here's the way some great historical moments might have flattened into nothingness, as reported in the press of the time.

The Israelite Daily Shofar, 1313 B.C.:
Egyptian authorities were snickering last night over

the Israelites' plan to escape through the Red Sea. "Our printouts confirm that the Red Sea is wet all the way to the bottom," giggled Anwar Kabat, a spokesman for the Pharaoh's government. "Moses will never make it."

Shortly before midnight Moses made a brief announcement to the press. "I am canceling the Exodus," he said. "Apparently the Lord made some serious mistakes in His computations. Everyone be back at work on the pyramids at six."

The Paris Inquisitor, 1431:
Authorities were relieved this week when a peasant girl named Joan renounced heretical remarks she had made earlier. Joan appeared before the Dauphin and said she had heard voices telling her to lead the French army to victory over the English. The girl refused to change her story even when threatened by fire and dismemberment. But she quickly recanted when the cathedral computer proved that her "voices" were, in all probability, the result of becoming disoriented by a Swiss bell-ringing contest held in her village. Joan was still burned at the stake, of course, but "without prejudice."

The Madrid Bullthrower, 1493:
Captain Christopher Columbus returned to Spain, redfaced, after giving up on his attempt to sail west to the Indies. "I must have been off my gourd," he said. "The computer became so hysterical when we got approximately fourteen leagues from the edge that I managed to come to my senses and turn back."

The San Antonio Spur, 1836:
Davy Crockett, smarting under allegations that he is "the coward of the Alamo," defended his actions in recommending that the Texans surrender to General Santa Anna. "It wasn't easy, being in charge of a computer that keeps telling you the Mexicans would win, the Alamo would be razed, we would all be forgotten, and something called a Tastee Freeze Taco stand would be built here," Crockett said. Meanwhile General Sam Houston, on hearing the news, disbanded his army and advised his troops to stop saying "greaser" and start learning Spanish.

The Munich Flügelhorn, 1890:
Eleven-year-old Albert Einstein was transferred from an academic course to a cooks' school, at the insistence of Professor Friedrich Platzhammer. "He's a nice boy," Platzhammer said, "but a mathematical dummkopf. He barely passes our computerized, multiple-choice tests, and he has the worst memory I've ever come across. He spends his days dreaming and doodling meaningless formulas. We fed it into the computer, and it blew a fuse. I just hope they can teach him to make strudel."

The London Tattler, 1893:
George Bernard Shaw burned his unpublished works in the middle of Fleet Street today, shouting that he would join the Church of England and go to America as a missionary. "I've no future in literature or drama," Shaw cried. "I've written five novels and three plays, and they have all been

banned or panned or both. The computer confirms that my stuff is all talk and no action. I'll never make it.''

MORE FANTASTIC READING!

THE WARLORD (1189, $3.50)
by Jason Frost
California has been isolated by a series of natural disasters. Now, only one man is fit to lead the people. Raised among Indians and trained by the Marines, Erik Ravensmith is a deadly adversary—and a hero of our times!

ORON #5: THE GHOST ARMY (1211, $2.75)
When a crazed tyrant and his army comes upon a village to sate their lusts, Oron stands between the warmonger and satisfaction—and only a finely-honed blade stands between Oron and death!

ORON (994, $1.95)
Oron, the intrepid warrior, joins forces with Amrik, the Bull Man, to conquer and rule the world. Science fantasy at its best!

THE SORCERER'S SHADOW (1025, $2.50)
In a science fantasy of swords, sorcery and magic, Akram battles an ageless curse. He must slay Attluma's immortal sorceress before he is destroyed by her love.

ORON: THE VALLEY OF OGRUM (1058, $2.50)
When songs of praise for Oron reach King Ogrum's ears, Ogrum summons a power of sorcery so fearsome that Oron's mighty broadsword may melt into a useless lump of steel!

MORE FANTASTIC READING FROM ZEBRA!

GONJI #1: DEATHWIND OF VEDUN   (1006, $3.25)
by T. C. Rypel
Cast out from his Japanese homeland, Gonji journeys across barbaric Europe in quest of Vedun, the distant city in the loftiest peaks of the Alps. Brandishing his swords with fury and skill, he is determined to conquer his hardships and fulfill his destiny!

GONJI #2: SAMURAI STEEL   (1072, $3.25)
by T. C. Rypel
His journey to Vedun is ended, but Gonji's most treacherous battle ever is about to begin. The invincible King Klann has occupied Vedun with his hordes of murderous soldiers—and plotted the samurai's destruction!

GONJI #3: SAMURAI COMBAT   (1191, $3.50)
by T. C. Rypel
King Klann and the malevolent sorcerer Mord have vanquished the city of Vedun and lay in wait to snare the legendary warrior Gonji. But Gonji dares not waiver—for to falter would seal the destruction of Vedun with the crushing fury of SAMURAI COMBAT!

SURVIVORS   (1071, $3.25)
by John Nahmlos
It would take more than courage and skill, more than ammo and guns, for Colonel Jack Dawson to survive the advancing nuclear war. It was the ultimate test—protecting his loved ones, defending his country, and rebuilding a civilization out of the ashes of war-ravaged America!

THE SWORD OF HACHIMAN   (1104, $3.50)
by Lynn Guest
Destiny returned the powerful sword of Hachiman to mighty Samurai warrior Yoshitsune so he could avenge his father's brutal death. Only he was unaware his most perilous enemy would be his own flesh and blood!

Available wherever paperbacks are sold, or order direct from the Publisher. Send cover price plus 50¢ per copy for mailing and handling to Zebra Books, 475 Park Avenue South, New York, N.Y. 10016 DO NOT SEND CASH.